Bocconi & Springer Series

Mathematics, Statistics, Finance and Economics

Volume 10

The **Bocconi & Springer Series** aims to publish research monographs and advanced textbooks covering a wide variety of topics in the fields of mathematics, statistics, finance, economics and financial economics. Concerning textbooks, the focus is to provide an educational core at a typical Master's degree level, publishing books and also offering extra material that can be used by teachers, students and researchers.

The series is born in cooperation with Bocconi University Press, the publishing house of the famous academy, the first Italian university to grant a degree in economics, and which today enjoys international recognition in business, economics, and law.

The series is managed by an international scientific Editorial Board. Each member of the Board is a top level researcher in his field, well-known at a local and global scale. Some of the Board Editors are also Springer authors and/or Bocconi high level representatives. They all have in common a unique passion for higher, specific education, and for books.

Volumes of the series are indexed in Web of Science - Thomson Reuters.

Manuscripts should be submitted electronically to Springer's mathematics editorial department: francesca.bonadei@springer.com

THE SERIES IS INDEXED IN SCOPUS

Laurent Decreusefond

Selected Topics in Malliavin Calculus

Chaos, Divergence and So Much More

BOCCONI
UNIVERSITY
PRESS

Laurent Decreusefond (iD)
Telecom Paris
Institut Polytechnique de Paris
Paris, France

ISSN 2039-1471   ISSN 2039-148X   (electronic)
Bocconi & Springer Series
ISBN 978-3-031-01313-3   ISBN 978-3-031-01311-9   (eBook)
https://doi.org/10.1007/978-3-031-01311-9

This Springer imprint is published by the registered company Springer Nature Switzerland AG
The registered company address is: Gewerbestrasse 11, 6330 Cham, Switzerland

To Sandrine, Marc, and Benjamin

Preface

It is sometimes easier to describe something by what it is not rather than by what it is supposed to be. This book is not a research monograph about Malliavin calculus with the latest results and the most sophisticated proofs. It does not contain all the known results even for the basic subjects addressed here. The goal was to give the largest possible variety of proof techniques. For instance, we did not focus on the proof of concentration inequality for functionals of the Brownian motion, as it follows closely the lines of the analog result for Poisson functionals.

This book grew from the graduate courses I gave at Paris-Sorbonne and Paris-Saclay universities, during the last few years. It is supposed to be as accessible as possible for students who have a knowledge of Itô calculus and some rudiments of functional analysis.

A recurrent difficulty when someone discovers Malliavin calculus is due to the different and often implicit identifications which are made between several functional spaces. I tried to demystify this point as much as possible. The presentation is hopefully self-contained, meaning that the necessary results of functional analysis which are supposed to be known in all the research monographs are recalled in the core of the text. The choice of the topics has been influenced by my own research which revolved for a while around fractional Brownian motion and then shifted to point processes, with an inclination to Stein's method.

I did not insist on the historical applications of the Malliavin calculus which were about the existence of the density of the distribution of some random variables, because there are so many other interesting subjects where the Malliavin calculus can be applied: Greeks computations, conditional expectations, change of measure, optimal transport, filtration enlargement, and, more recently, the Stein-Malliavin method.

I am greatly indebted to A.S. Üstünel who introduced me to Malliavin calculus a few years ago. It has been a long and rich journey since then.

This book benefited from the help of numerous students, most notably B. Costacèque-Cecchi. The remaining errors are mine.

Paris, France   Laurent Decreusefond
2021

Contents

About the Author

Laurent Decreusefond is a former student of Ecole Normale Supérieure de Paris-Saclay. He received the Agrégation in 1989, his PhD in 1994, and his habilitation in 2001 in mathematics. He is now Full Professor of Mathematics at Institut Polytechnique de Paris, one of the most renowned French research and teaching institutions.

His research topics are twofold. The theoretical part is devoted to Malliavin calculus and its applications. He is the author of a highly cited paper about fractional Brownian motion that paved the way to a thousand research articles. Recently, he has been interested in the functional Stein-Malliavin method, which gives the convergence rate in functional limit theorems.

On a more applied part, he proposed new paradigms for stochastic modelling of telecom systems, including stochastic geometry and random topological algebra.

He coauthored several papers that gave a new approach to the coverage analysis of cellular systems. The performance of some of the algorithms so defined may be analyzed with mathematical tools coming from the Malliavin calculus, such as concentration inequalities.

Chapter 1
Wiener Space

Abstract The construction and the characterization of probability measures on infinite dimensional spaces is a hard task. Gaussian random variables and vectors have multiple properties which permit to construct Gaussian measures on Banach spaces. This procedure yields the notion of Abstract Wiener Space of Gelfand Triplet.

1.1 Gaussian Random Variables

We begin by basic definitions about Gaussian random variables and vectors.

Definition 1.1 (Gaussian Random Variable) A real-valued random variable X is Gaussian whenever its characteristic function is of the form

$$\mathbf{E}\left[e^{itX}\right] = e^{itm}e^{-\sigma^2t^2/2}. \tag{1.1}$$

It is well known that $\mathbf{E}[X] = m$ and $\mathrm{Var}(X) = \sigma^2$.

Remark 1.1 This definition means that whenever we know that a random variable is Gaussian, it is sufficient to compute its average and its variance to fully determine its distribution.

A Gaussian random vector is not simply a collection of Gaussian random variables. It is true that all the coordinates of a Gaussian vector are Gaussian, but they do satisfy a supplementary condition. In what follows, the Euclidean scalar product on $\mathbf{R}^n$ is defined by

$$\langle x,\, y \rangle = \sum_{j=1}^{n} x_j y_j.$$

Definition 1.2 (Gaussian Random Vector) A random vector X in $\mathbf{R}^n$, i.e., $X = (X_1, \cdots, X_n)$, is a Gaussian random vector whenever for any $t = (t_1, \cdots, t_n) \in$

1
L. Decreusefond, *Selected Topics in Malliavin Calculus*,
Bocconi & Springer Series 10, https://doi.org/10.1007/978-3-031-01311-9_1

$\mathbf{R}^n$, the real-valued random variable

$$\langle t,\, X \rangle = \sum_{j=1}^{n} t_j X_j$$

is Gaussian.

In view of the Remark 1.1, we have

$$\mathbf{E}\left[e^{i\langle t,\, X \rangle} \right] = e^{i\langle t,\, m \rangle} e^{-\frac{1}{2}\langle \Gamma_X t,\, t \rangle}, \tag{1.2}$$

where

$$\Gamma_X = \left(\mathrm{cov}(X_j,\, X_k),\, 1 \le j, k \le n \right)$$

is the so-called covariance matrix of X.

Remark 1.2 Somehow hidden in the previous definition lies the identity

$$\mathrm{Var}\,\langle t,\, X \rangle = \sum_{i,j=1}^{n} \mathrm{cov}(X_j,\, X_k) t_i t_j \tag{1.3}$$

for any $t = (t_1, \cdots, t_n) \in \mathbf{R}^n$. Since a variance is always non-negative, this means that Γ_X satisfies the identity

$$\langle \Gamma_X t,\, t \rangle = \sum_{i,j=1}^{n} \Gamma_X(i,\, j)\, t_i\, t_j \ge 0,$$

which induces that the eigenvalues of Γ_X are non-negative.

The main feature of Gaussian vectors is that they are stable by affine transformation.

Theorem 1.1 *Let X be an $\mathbf{R}^n$-valued Gaussian vector, $B \in \mathbf{R}^p$ and A a linear map (i.e., a matrix) from $\mathbf{R}^n$ into $\mathbf{R}^p$. The random $Y = AX + B$ is an $\mathbf{R}^p$-valued Gaussian vector whose characteristics are given by*

$$\mathbf{E}[Y] = A\mathbf{E}[X] + B, \quad \Gamma_Y = A\Gamma_X A^t,$$

where A^t is the transpose of A.

Remark 1.3 For X, a one dimensional centered, Gaussian random variable,

$$\mathbf{E}\left[|X|^p \right] = c_p\, \mathrm{Var}(X)^{p/2}. \tag{1.4}$$

Actually, in view of the previous theorem,

$$\mathbf{E}\left[|\mathcal{N}(0, \sigma^2)|^p\right] = \sigma^{p/2}\mathbf{E}\left[|\mathcal{N}(0, 1)|^p\right].$$

Remark 1.4 If Γ is non-negative symmetric matrix, one can define $\Gamma^{1/2}$, a symmetric non-negative matrix whose square equals Γ. If $X = (X_1, \cdots, X_n)$ is a vector of independent standard Gaussian random variables, then the previous theorem entails that $\Gamma^{1/2}X$ is a Gaussian vector of covariance matrix Γ.

Beyond this stability by affine transformation, the set of Gaussian vectors enjoys another remarkable stability property.

Theorem 1.2 *Let* $(X_n, n \geq 1)$ *be a sequence of Gaussian vectors,* $X_n \sim \mathcal{N}(m_n, \Gamma_{X_n})$, *which converges in distribution to some random vector* X. *Then,* X *is a Gaussian vector* $\mathcal{N}(m, \Gamma_X)$ *where*

$$m = \lim_{n\to\infty} m_n \text{ and } \Gamma_X = \lim_{n\to\infty} \Gamma_{X_n}.$$

Remark that for $X \sim \mathcal{N}(0, \mathrm{Id}_n)$, a standard Gaussian vector in $\mathbf{R}^n$,

$$\mathbf{E}\left[\|X\|_{\mathbf{R}^n}^2\right] = \sum_{j=1}^{n} \mathbf{E}\left[X_j^2\right] = n.$$

This means that the mean norm of such a random variable goes to infinity as the dimension grows. Thus, we cannot construct a Gaussian distribution on an infinite dimensional space like $\mathbf{R}^{\mathbf{N}}$, by just extending what we do on $\mathbf{R}^n$.

Definition 1.3 (Gaussian Processes) For a set T, a family $(X(t), t \in T)$ of random variables is a Gaussian process whenever for any $n \geq 1$, for any $(t_1, \cdots, t_n) \in T^n$, the random vector $(X(t_1), \cdots, X(t_n))$ is a Gaussian vector.

1.2 Wiener Measure

The construction of measures on functional spaces is a delicate question that is satisfactorily solved for Gaussian measures. Recall that a Brownian motion is defined as follows.

Definition 1.4 The Brownian motion $B = (B(t), t \geq 0)$ is the (unique) centered, Gaussian process on $\mathbf{R}^+$ with independent increments such that

$$\mathbf{E}[B(t)B(s)] = t \wedge s.$$

Its sample-paths are Hölder continuous of any order strictly less than $1/2$.

As such, the distribution of B defines a measure on the space of continuous functions, null at time 0, as well as a measure on the spaces $\mathrm{Hol}(\alpha)$ for any $\alpha < 1/2$. It remains to prove that such a process does exist. There are several possibilities to do so. The most intuitive is probably the Donsker–Lamperti theorem:

Theorem 1.3 (Donsker–Lamperti) *Let $(X_n, n \geq 1)$ be a sequence of independent, identically distributed random variables such that $\mathbf{E}\left[|X_1|^{2p}\right] < \infty$. Then,*

$$\frac{1}{\sqrt{n}} \sum_{j=1}^{[nt]} X_j \Longrightarrow B(t)$$

in the topology of $\mathrm{Hol}(\gamma)$ *for any* $\gamma < (p-1)/2p$, *i.e.,*

$$\mathbf{E}\left[F\left(\frac{1}{\sqrt{n}} \sum_{j=1}^{[n.]} X_j\right) \right] \xrightarrow{n \to \infty} \mathbf{E}\left[F(B)\right]$$

for any $F^{\cdot}: \mathrm{Hol}(\gamma) \to \mathbf{R}$ bounded and continuous.

For $p = 1$, i.e., square integrable random variables, the convergence holds in $C([0, T]; \mathbf{R})$ for any $T > 0$.

This construction of the Brownian motion via the random walk is not fully satisfactory as we cannot write B as the sum of a series. The construction of Itô-Nisio is more interesting in this respect.

We need to introduce a few functional spaces before going further. The most well known space of functions is the set of continuous functions.

Definition 1.5 (Space of Continuous Functions) We denote by C the space of real-valued functions, continuous on $[0, 1]$, null at time 0 equipped with the norm

$$\|f\|_\infty = \sup_{t \in [0,1]} |f(t)|.$$

The space C is a complete normed space, i.e., a Banach space. The polynomials are dense in C hence it is separable.

If we look at further properties of functions, there are a multitude of ways a function can be *more than* continuous but not differentiable. This means that there exists a bunch of spaces between C^1 and C. The most celebrated are probably the Hölder spaces.

Definition 1.6 (Hölder Space) For $\alpha \in (0, 1]$, a function $f : [0, 1] \to \mathbf{R}$ is said to be Hölder continuous of order α whenever there exists $c > 0$ such that for all $s, t \in [0, 1]$,

$$|f(t) - f(s)| \leq c|t - s|^{\alpha}.$$

The norm on $\mathrm{Hol}(\alpha)$ is given by

$$\|f\|_{\mathrm{Hol}(\alpha)} = |f(0)| + \sup_{s \neq t} \frac{|f(t) - f(s)|}{|t - s|^\alpha}.$$

With this norm, $\mathrm{Hol}(\alpha)$ is a Banach space, but it is not separable. When $\alpha = 1$, the functions are said to be Lipschitz continuous.

Remark 1.5 In what follows ℓ denotes the Lebesgue measure on $\mathbf{R}$ or $\mathbf{R}^n$ according to the context.

Alternatively, we may consider Sobolev like spaces that are often easier to work with despite their apparent complexity.

Definition 1.7 (Riemann–Liouville Fractional Spaces) For $\alpha > 0$, for $f \in L^2([0, 1] \to \mathbf{R}; \ell)$,

$$I^\alpha f(t) = \frac{1}{\Gamma(\alpha)} \int_0^t (t - s)^{\alpha - 1} f(s) \mathrm{d}s. \tag{1.5}$$

The space $I_{\alpha,2}$ is the set $I^\alpha(L^2([0, 1] \to \mathbf{R}; \ell))$ equipped with the scalar product

$$\langle I^\alpha f, I^\alpha g \rangle_{I_{\alpha,2}} = \langle f, g \rangle_{L^2([0,1] \to \mathbf{R};\, \ell)} = \int_0^1 f(s) g(s) \mathrm{d}s.$$

Since the map $(f \mapsto I^\alpha f)$ is one-to-one, this defines a scalar product.

   More generally, for $p \geq 1$, $I_{\alpha,p}$ is the space $I^\alpha(L^p([0, 1] \to \mathbf{R}; \ell))$ equipped with the norm

$$\|I^\alpha f\|_{I_{\alpha,p}} = \|f\|_{L^p([0,1] \to \mathbf{R};\, \ell)}.$$

Another useful scale of functions is the Slobodetzky family of fractional Sobolev spaces.

Definition 1.8 (Slobodetzky Spaces) For $\alpha \in (0, 1]$ and $p \geq 1$, a function $f \in L^p([0, 1] \to \mathbf{R}; \ell)$ is in $W_{\alpha,p}$ whenever

$$\iint_{[0,1]^2} \frac{|f(t) - f(s)|^p}{|t - s|^{1+\alpha p}} \mathrm{d}s \mathrm{d}t < \infty.$$

The space $W_{\alpha,p}$ equipped with the norm

$$\|f\|_{W_{\alpha,p}}^p := \|f\|_{L^p([0,1] \to \mathbf{R};\, \ell)} + \left(\iint_{[0,1]^2} \frac{|f(t) - f(s)|^p}{|t - s|^{1+\alpha p}} \mathrm{d}s \mathrm{d}t \right)^{1/p},$$

is a separable Banach space.

These spaces are interesting because of the following embeddings.

Theorem 1.4 *For any $\alpha'' > \alpha' > \alpha > 1/p$, we have*

$$\text{Hol}(\alpha'') \subset W_{\alpha',p} \subset I_{\alpha,p} \subset \text{Hol}(\alpha - 1/p) \subset C.$$

Moreover, polynomials on $[0, 1]$ have bounded derivative, thus they are Lipschitz hence Hölder continuous of any order and they are dense in C hence all these spaces are dense in C.

As a consequence, we retrieve easily the Kolmogorov lemma about the regularity of Brownian sample-paths.

Lemma 1.1 *For any $\alpha \in [0, 1/2)$ and any $p \geq 1$, the sample-paths of a Brownian motion belong to $W_{\alpha,p}$ with probability 1.*

Proof It is sufficient to prove that

$$\mathbf{E}\left[\iint_{[0,1]^2} \frac{|B(t) - B(s)|^p}{|t - s|^{1+\alpha p}} ds dt\right] < \infty.$$

Since $B(t) - B(s)$ is a Gaussian random variable,

$$\mathbf{E}\left[|B(t) - B(s)|^p\right] = c_p \mathbf{E}\left[|B(t) - B(s)|^2\right]^{p/2} = c_p|t - s|^{p/2}.$$

The function $(s, t) \longmapsto |t - s|^{-1+(1/2-\alpha)p}$ is integrable provided that $\alpha < 1/2$, hence the result.   □

A space which will be of paramount importance in the following is the Cameron–Martin space.

Definition 1.9 (Cameron–Martin Space) The Cameron–Martin space, denoted by $\mathcal{H}$, is defined by $\mathcal{H} = I_{1,2}$, the set of differentiable functions whose derivative is square integrable over $[0, 1]$, equipped with the scalar product

$$\langle f, g \rangle_{\mathcal{H}} = \langle \dot{f}, \dot{g} \rangle_{L^2\left([0,1] \to \mathbf{R}; \ell\right)}$$

where $\dot{f}$ is the unique element of $L^2([0, 1] \to \mathbf{R}; \ell)$ such that

$$f(t) = I^1 f(t) = \int_0^t \dot{f}(s) ds.$$

According to the Cauchy–Schwarz inequality, for $f \in \mathcal{H}$

$$|f(t) - f(s)| = \left|\int_0^1 \mathbf{1}_{(s,t]}(r) \dot{f}(r) dr\right| \leq \sqrt{t - s} \, \|\dot{f}\|_{L^2\left([0,1] \to \mathbf{R}; \ell\right)}$$

hence $\mathcal{H} \subset \text{Hol}(1/2)$ and in view of Theorem 1.4, $\mathcal{H}$ is dense in any $W_{\alpha,p}$ for $\alpha < 1/2$, $p \geq 1$.

We now are in position to describe the Itô-Nisio construction of the Wiener measure. Consider $(\dot{h}_m, m \geq 0)$ a complete orthonormal basis of $L^2([0,1] \to \mathbf{R}; \ell)$. By the very definition of the scalar product on $\mathcal{H}$, this entails that $(h_m = I^1\dot{h}_m, m \geq 0)$ is a complete orthonormal basis of $\mathcal{H}$. One may choose the family given by

$$h_0(t) = t \text{ and } h_m(t) = \frac{\sqrt{2}}{\pi m} \sin(\pi m t) \text{ for } m \geq 1. \tag{1.6}$$

Then, consider the sequence of approximations given by

$$S_n(t) = \sum_{m=0}^{n} X_m h_m(t) \tag{1.7}$$

where $(X_m, m \geq 0)$ is a sequence of independent, standard Gaussian random variables. We then have the following extension of the Itô-Nisio theorem.

Theorem 1.5 *For any (α, p) such that $1/p < \alpha < 1/2$, the sequence $(S_n, n \geq 1)$ converges in $W_{\alpha,p}$ with probability 1. Moreover, the limit process, denoted by B, is Gaussian, centered with covariance*

$$\mathbf{E}[B(t)B(s)] = \min(t,s).$$

Hence B is distributed as a Brownian motion (Fig. 1.1).

We first need a general lemma.

Lemma 1.2 *Let*

$$\omega_M = \sup_{m,n \geq M} \|S_n - S_m\|_{W_{\alpha,p}} \text{ and } T_M = \sup_{n \geq M} \|S_n - S_M\|_{W_{\alpha,p}}.$$

Fig. 1.1 A sample-path of S_{5000} sampled on one thousand points. The roughness is already apparent though the trajectory is still differentiable

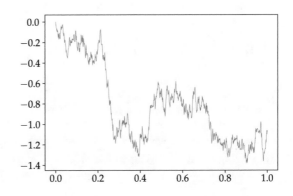

If $(T_M, M \geq 1)$ converges in probability to 0, then $(S_n, n \geq 1)$ is convergent with probability 1.

Proof It is clear that

$$\left(T_M \leq \epsilon\right) \subset \left(\omega_M \leq 2\epsilon\right),$$

hence

$$\mathbf{P}(\omega_M > 2\epsilon) \leq \mathbf{P}(T_M > \epsilon).$$

If $(T_M, M \geq 1)$ converges in probability to 0, then so does $(\omega_M, M \geq 1)$. Consequently, there is a subsequence that converges with probability 1 but ω_M is decreasing, hence the whole sequence $(\omega_M, M \geq 1)$ converges to 0 with probability 1.

This means that $(S_n, n \geq 1)$ is a.e. a Cauchy sequence in a complete Banach space, hence is convergent.   □

Proof of Theorem 1.5 **Step 1**   The Doob inequality for Banach valued martingales states that

$$\mathbf{E}\left[T_M^p\right] \leq \frac{p}{p-1} \sup_{n \geq M} \mathbf{E}\left[\|S_n - S_M\|_{W_{\alpha,p}}^p\right] \tag{1.8}$$

Since $S_n - S_M$ is a Gaussian process, in view of (1.4),

$$\mathbf{E}\left[\left|(S_n - S_M)(t) - (S_n - S_M)(s)\right|^p\right]$$

$$= c_p \mathbf{E}\left[\left|(S_n - S_M)(t) - (S_n - S_M)(s)\right|^2\right]^{p/2}$$

$$= c_p \mathbf{E}\left[\left(\sum_{m=M+1}^{n} X_m \left(h_m(t) - h_m(s)\right)\right)^2\right]^{p/2}.$$

Since the X_m's are independent with unit variance,

$$\mathbf{E}\left[\left(\sum_{m=M+1}^{n} X_m \left(h_m(t) - h_m(s)\right)\right)^2\right] = \sum_{m=M+1}^{n} \left(h_m(t) - h_m(s)\right)^2. \tag{1.9}$$

A Clever Use of Parseval Identity
The trick is to note that

$$h_m(t) = \langle \dot{h}_m, \mathbf{1}_{[0,t]} \rangle_{L^2} = \langle h_m, t \wedge . \rangle_{\mathcal{H}}.$$

□

This means that the right-hand side of (1.9) is the Cauchy remainder of the series

$$\sum_{m=0}^{\infty} \langle h_m, t \wedge . - s \wedge . \rangle_{\mathcal{H}}^2 = \|t \wedge . - s \wedge . \|_{\mathcal{H}}^2 = |t - s|,$$

according to the Parseval identity. Since $\alpha < 1/2$,

$$\int_{[0,1]^2} |t - s|^{p/2} |t - s|^{-1-\alpha p} ds dt = \int_{[0,1]^2} |t - s|^{-1+(1/2-\alpha)p} ds dt < \infty.$$

Similarly, we have

$$\mathbf{E}\left[\|S_n - S_M\|_{L^p\left([0,1] \to \mathbf{R}; \ell\right)}^p \right] \le c \left(\sum_{m=M+1}^{n} \langle h_m, t \wedge . \rangle_{\mathcal{H}}^2 \right)^{p/2}.$$

By the dominated convergence theorem, it follows that

$$\sup_{n \ge M} \mathbf{E}\left[\|S_n - S_M\|_{W_{\alpha,p}}^p \right] \le c \int_0^1 \left(\sum_{m=M+1}^{\infty} \langle h_m, t \wedge . \rangle_{\mathcal{H}}^2 \right)^{p/2} dt$$

$$+ c \int_{[0,1]^2} \left(\sum_{m=M+1}^{\infty} \langle h_m, t \wedge . - s \wedge . \rangle_{\mathcal{H}}^2 \right)^{p/2} |t - s|^{-1-\alpha p} ds dt$$

$$\xrightarrow{M \to \infty} 0. \qquad (1.10)$$

The result follows from (1.10), the Markov inequality, and Lemma 1.2. We denote by B the limit of S_n.

Step 2 It is clear that for any $(t_1, \cdots, t_n) \in [0, 1]$ and $(\alpha_1, \cdots, \alpha_n)$,

$$\sum_{i=1}^{n} \alpha_i S_M(t_i)$$

is a Gaussian random variable. In view of Theorem 1.2, the limit is Gaussian hence B is a Gaussian process.

Step 3 Remark that the sequence $(S_n, n \ge 1)$ is built on the probability space $\Omega = \mathbf{R}^{\mathbf{N}}$, equipped with the probability measure $\mathbf{P} = \otimes_{n \in \mathbf{N}} \nu$ where ν is the standard Gaussian distribution on $\mathbf{R}$. Fatou's Lemma and (1.9) entail that

$$\mathbf{E}\left[\|B - S_M\|_{W_{\alpha,p}}^p \right] \le \liminf_n \mathbf{E}\left[\|S_n - S_M\|_{W_{\alpha,p}}^p \right]$$

$$\le \limsup_n \mathbf{E}\left[\|S_n - S_M\|_{W_{\alpha,p}}^p \right]$$

$$= \inf_{M} \sup_{n \geq M} \mathbf{E}\left[\|S_n - S_M\|^p_{W_{\alpha,p}}\right]$$

$$= 0, \tag{1.11}$$

according to (1.10). This means $(S_M, M \geq 1)$ converges to B in $L^2(\Omega \to W_{\alpha,2}; \mathbf{P})$, hence

$$\mathbf{E}[B(t)B(s)] = \mathbf{E}\left[\sum_{m=0}^{\infty} X_m \langle h_m, t \wedge . \rangle_{\mathcal{H}} \times \sum_{m'=0}^{\infty} X_{m'} \langle h_{m'}, s \wedge . \rangle_{\mathcal{H}}\right]$$

$$= \sum_{m=0}^{\infty} \sum_{m'=0}^{\infty} \mathbf{E}[X_m X_{m'}] \langle h_m, t \wedge . \rangle_{\mathcal{H}} \langle h_{m'}, s \wedge . \rangle_{\mathcal{H}}$$

by Fubini Theorem,

$$= \sum_{m=0}^{\infty} \mathbf{E}\left[X_m^2\right] \langle h_m, t \wedge . \rangle_{\mathcal{H}} \langle h_m, s \wedge . \rangle_{\mathcal{H}}$$

by independence and hence orthogonality of the X_m's,

$$= \sum_{m=0}^{\infty} \langle h_m, t \wedge . \rangle_{\mathcal{H}} \langle h_m, s \wedge . \rangle_{\mathcal{H}}$$

since X_m has a unit variance,

$$= \langle t \wedge ., s \wedge . \rangle_{\mathcal{H}},$$

according to the Parseval equality. The very definition of the scalar product on $\mathcal{H}$ entails that

$$\langle t \wedge ., s \wedge . \rangle_{\mathcal{H}} = \int_0^1 \mathbf{1}_{[0,t]}(r)\mathbf{1}_{[0,s]}(r)\mathrm{d}r = t \wedge s.$$

Several other constructions as limit of stochastic processes lead to a Brownian motion. As a conclusion of these theorems, it appears that the distribution of B is a probability measure on the Banach spaces $C([0, 1]; \mathbf{R})$, $\mathrm{Hol}(\gamma)$ or $W_{\alpha,p}$. Now, if we reverse the problem, how can we characterize a probability measure on, say, $C([0, 1]; \mathbf{R})$? How do we determine that it coincides with the Brownian motion distribution?

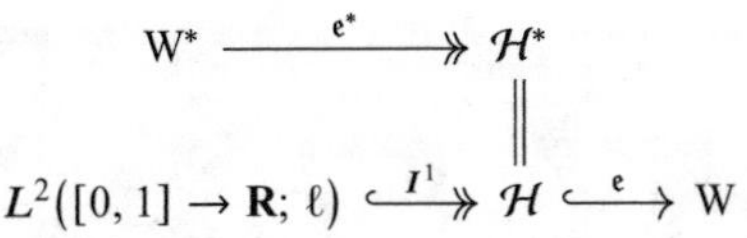

$$W^* \xrightarrow{\;\;\mathfrak{e}^*\;\;} \mathcal{H}^*$$

$$\|$$

$$L^2([0,1] \to \mathbf{R}; \ell) \overset{I^1}{\hookrightarrow\kern-0.8em\twoheadrightarrow} \mathcal{H} \overset{\mathfrak{e}}{\hookrightarrow} W$$

Fig. 1.2 Embeddings and identification for Wiener spaces. An arrow with a hook means the map is one-to-one. A double head indicates that the map is onto or that is range in dense

In finite dimension, a probability measure is characterized by its Fourier transform, often called its characteristic function. This still holds in separable Banach spaces.

Definition 1.10 For μ a probability measure on a separable Banach space W (whose dual is denoted by W^*), its characteristic functional is

$$\phi_\mu : W^* \longrightarrow \mathbf{C}$$

$$z \longmapsto \int_W e^{i\langle z,w\rangle_{W^*,W}} d\mu(w).$$

Theorem 1.6 *For μ and ν two probability measures on W,*

$$(\phi_\mu = \phi_\nu) \Longleftrightarrow (\mu = \nu).$$

We now need to introduce the set of functional spaces that will serve as the framework for the sequel. From now on, W will be any of the spaces $W_{\alpha,p}$ for $1/p < \alpha < 1/2$ or $C([0,1], \mathbf{R})$ and W^* is its topological dual (the set of *continuous* linear forms on W). The measure μ is the Wiener measure, i.e., the distribution induced by the Brownian motion on W.

The Hilbert space $\mathcal{H}$ plays the rôle of pivotal space, meaning that it is identified with its dual. The map $\mathfrak{e}$ is the embedding from $\mathcal{H}$ into W and $\mathfrak{e}^*$ is its adjoint map. Because of the identification, we have that for any $z \in W^*$ and $h \in \mathcal{H}$,

$$\langle z, \mathfrak{e}(h)\rangle_{W^*,W} = \langle \mathfrak{e}^*(z), h\rangle_{\mathcal{H}}.$$

It is useful to have in mind the diagram of Fig. 1.2.

Note that, since $\mathfrak{e}^*(W^*)^\perp = \ker \mathfrak{e} = \{0\}$, $\mathfrak{e}^*(W^*)$ is dense in $\mathcal{H}$. The triplet $(W, \mathcal{H}, \mu)$ is known as a Gelfand triplet or an abstract Wiener space.

Do Not Identify Too Much!
As the map I^1, the first order quadrature operator, is an isometric isomorphism between $L^2([0,1] \to \mathbf{R}; \ell)$ and $\mathcal{H}$, it is common to identify these two spaces. Since we already identified $\mathcal{H}$ and its dual, we cannot identify $\mathcal{H}$

(continued)

and $L^2([0, 1] \to \mathbf{R}; \ell)$ otherwise $\mathcal{H}$ is identified to $L^2([0, 1] \to \mathbf{R}; \ell)$, i.e., all square integrable functions are differentiable. Unfortunately, this is often done in the literature because it simplifies greatly the presentation and permits useful further identifications. This is the main cause of disarray when first reading a paper or a book on Malliavin calculus.

Example (Representation of $\mathfrak{e}^*(\varepsilon_a)$*)* According to Theorem 1.4, $\mathcal{H} \subset \mathrm{Hol}(1/2)$. Thus, the Dirac measure ε_a is a continuous linear map on $\mathcal{H}$. Let x_a be its representative in $\mathcal{H}$. We must have for any $f \in \mathcal{H}$,

$$\varepsilon_a(f) = f(a) = f(a) - f(0) = \langle x_a, f \rangle_{\mathcal{H}} = \int_0^1 \dot{x}_a(s) \dot{f}(s) \, ds,$$

where $\dot{f} = (I^1)^{-1} f$ is the derivative of f. The sole candidate is $\dot{x}_a = \mathbf{1}_{[0,a]}$, hence $x_a(s) = a \wedge s$, i.e.,

$$\mathfrak{e}^*(\varepsilon_a) = . \wedge a. \tag{1.12}$$

Hence, $\mathfrak{e}^*(\varepsilon_a) = a \wedge . = I^1(\mathbf{1}_{[0,a]})$.

With the notations of Theorem 1.5, we have

Theorem 1.7 *For any* $z \in \mathrm{W}^*$,

$$\mathbf{E}\left[e^{i \langle z, B \rangle_{\mathrm{W}^*, \mathrm{W}}}\right] = \exp\left(-\frac{1}{2} \|\mathfrak{e}^*(z)\|_{\mathcal{H}}^2\right). \tag{1.13}$$

Proof From Theorem 1.5, we have

$$\langle z, B \rangle_{\mathrm{W}^*, \mathrm{W}} = \lim_{n \to \infty} \sum_{m=0}^n X_m \langle z, \mathfrak{e}(h_m) \rangle_{\mathrm{W}^*, \mathrm{W}}.$$

Remark that the random variable $\langle z, B \rangle_{\mathrm{W}^*, \mathrm{W}}$ is the limit of a sum of independent Gaussian random variables. By dominated convergence, we get

$$\mathbf{E}\left[e^{i \langle z, B \rangle_{\mathrm{W}^*, \mathrm{W}}}\right] = \lim_{n \to \infty} \prod_{m=0}^n \mathbf{E}\left[e^{i X_m \langle z, \mathfrak{e}(h_m) \rangle_{\mathrm{W}^*, \mathrm{W}}}\right]$$

$$= \lim_{n \to \infty} \prod_{m=0}^n \exp\left(-\frac{1}{2} \langle z, \mathfrak{e}(h_m) \rangle_{\mathrm{W}^*, \mathrm{W}}^2\right)$$

by (1.1),

$$= \exp\left(-\frac{1}{2}\sum_{m=0}^{\infty}\langle e^*(z), h_m\rangle^2_{W^*,W}\right)$$

$$= \exp\left(-\frac{1}{2}\|e^*(z)\|^2_{\mathcal{H}}\right),$$

according to the Parseval equality. □

1.3 Wiener Integral

The dual bracket between an element of W^* and an element of W is defined by construction of the dual of W. But we not only have the Banach structure on W, we also have a measure. We can take advantage of this richer framework to extend the above mentioned dual bracket to elements of $\mathcal{H}$ and W. In what follows, the letter ω represents the generic element of W. We denote by μ the distribution of B on W.

Definition 1.11 (Wiener Integral) The map

$$\delta : e^*(W^*) \subseteq \mathcal{H} \longrightarrow L^2(W \to \mathbf{R}; \mu)$$

$$e^*(z) \longmapsto \langle z, \omega\rangle_{W^*,W}.$$

is an isometry. Its unique extension to $\mathcal{H}$ is called the Wiener integral.

Proof The very definition of μ (see (1.13)) entails that for any $z \in W^*$,

$$\mathbf{E}\left[\exp\left(i\theta\,\langle z, \omega\rangle_{W^*,W}\right)\right] = \exp\left(-\frac{\theta^2}{2}\|e^*(z)\|^2_{\mathcal{H}}\right).$$

This means that the random variable $(\delta z)(\omega) = \langle z, \omega\rangle_{W^*,W}$ is a centered Gaussian random variable of variance $\|e^*(z)\|^2_{\mathcal{H}}$. Otherwise stated, for $h \in e^*(W^*)$,

$$\|\delta(h)\|_{L^2(W\to\mathbf{R};\,\mu)} = \|h\|_{\mathcal{H}}. \tag{1.14}$$

Since $e^*(W^*)$ is dense in $\mathcal{H}$, we can extend δ as a linear isometry from $\mathcal{H}$ into $L^2(W \to \mathbf{R}; \mu)$ as follows: For $h \in \mathcal{H}$, take $(z_n, n \geq 1)$ a sequence of elements of W^* such that $e^*(z_n)$ converges to $h \in \mathcal{H}$. Then according to (1.14), the sequence $(\delta(e^*(z_n)), n \geq 1)$ is Cauchy in $L^2(W \to \mathbf{R}; \mu)$ hence converges to an element of $L^2(W \to \mathbf{R}; \mu)$ we denote by δh. Moreover, (1.14) also implies that if $h = 0$ then $\lim_{n\to\infty} \delta(e^*(z_n)) = 0$ hence the limit does not depend on the chosen sequence. □

Corollary 1.1 *For $h \in \mathcal{H}$,*

$$\mathbf{E}\left[e^{i\delta(h)}\right] = \exp\left(-\frac{1}{2}\|h\|_{\mathcal{H}}^2\right).$$

Proof With our new notations, Eq. (1.13) can be rewritten as follows: For $z \in W^*$,

$$\mathbf{E}\left[e^{i\delta(\mathfrak{e}^*(z))}\right] = \exp\left(-\frac{1}{2}\|\mathfrak{e}^*(z)\|_{\mathcal{H}}^2\right).$$

Let $(z_n,\ n \geq 1)$ be a sequence of elements of W^* such that $\mathfrak{e}^*(z_n)$ tends to h in $\mathcal{H}$. By construction, $(\delta(\mathfrak{e}^*(z_n)),\ n \geq 1)$ tends to $\delta(h)$ in $L^2(W \to \mathbf{R};\ \mu)$, hence there is a subsequence that we still denote by the same indices, which converges with probability 1 in W. The dominated convergence theorem thus entails that

$$\mathbf{E}\left[e^{i\delta(\mathfrak{e}^*(z_n))}\right] \xrightarrow{n \to \infty} \mathbf{E}\left[e^{i\delta(h)}\right].$$

Furthermore, we have

$$\|\mathfrak{e}^*(z_n)\|_{\mathcal{H}} \xrightarrow{n \to \infty} \|h\|_{\mathcal{H}}$$

and the result follows.   □

Remark 1.6 For $h \in W^*$ and $k \in \mathcal{H}$

$$\langle h, \omega + \mathfrak{e}(k)\rangle_{W^*,W} = \delta(\mathfrak{e}^*(h))(\omega) + \langle \mathfrak{e}^*(h),\ k\rangle_{\mathcal{H}}.$$

Passing to the limit, we have

$$\delta h(\omega + k) = \delta h(\omega) + \langle h, k\rangle_{\mathcal{H}}, \tag{1.15}$$

for any $h \in \mathcal{H}$.

Remark 1.7 In view of (1.12), we can write

$$\omega(t) = \delta(\mathfrak{e}^*(\varepsilon_t)) = \delta(t \wedge .)(\omega). \tag{1.16}$$

Furthermore, let $(h_m,\ m \geq 0)$ be a complete orthonormal basis of $\mathcal{H}$. We have

$$t \wedge . = \sum_{m=0}^{\infty} \langle t \wedge .,\ h_m\rangle_{\mathcal{H}} h_m$$

$$= \sum_{m=0}^{\infty} \langle \mathbf{1}_{[0,t]},\ \dot{h}_m\rangle_{L^2([0,1]\to\mathbf{R};\ \ell)} h_m$$

$$= \sum_{m=0}^{\infty} h_m(t)\, h_m.$$

Hence,

$$\omega(t) = \sum_{m=0}^{\infty} \delta h_m(\omega)\, \mathfrak{e}(h_m)(t) \tag{1.17}$$

A Word About the Notations

The Brownian motion takes its value in W. We can see it as a random variable from an indefinite space Ω into W and then use the notation B, implicitly representing $B(\omega)$. With this convention, the distribution of B on W is the Wiener measure and denoted by μ. We can as well choose Ω to be itself W, i.e., work on what is called the canonical space, and then $B(\omega) = \omega$. In this situation, as usual the measure on Ω is denoted by **P**. Thus for $F : W \to \mathbf{R}$, we can equivalently write

$$\int_W F \mathrm{d}\mu \text{ or } \mathbf{E}\,[F(B)].$$

No notation is better than the other. The former is more usual, the latter keeps track of the fact that we are working with trajectories as basic elements.

The most useful theorem for the sequel states that if we translate the Brownian sample-path by an element of $\mathcal{H}$, then the distribution of this new process is absolutely continuous with respect to the initial Wiener measure. This is the transposition in infinite dimension of the trivial result in dimension 1:

$$\mathbf{E}\,[f(\mathcal{N}(m, 1))] = (2\pi)^{-1/2} \int_{\mathbf{R}} f(x + m)e^{-x^2/2}\mathrm{d}x$$

$$= (2\pi)^{-1/2} \int_{\mathbf{R}} f(x)e^{xm-m^2/2}e^{-x^2/2}\mathrm{d}x = \mathbf{E}\,[(f\Lambda_m)(\mathcal{N}(0, 1))]$$

where $\Lambda_m(x) = e^{xm-m^2/2}$.

Theorem 1.8 (Cameron–Martin) *For any $h \in \mathcal{H}$, for any bounded function $F :$ W $\to$ **R**,*

$$\int_W F(\omega + \mathfrak{e}(h))\mathrm{d}\mu(\omega) = \int_W F(\omega)\Lambda_h(\omega)\mathrm{d}\mu(\omega) \tag{1.18}$$

where

$$\Lambda_h(\omega) = \exp\left(\delta h(\omega) - \frac{1}{2}\|h\|_{\mathcal{H}}^2\right).$$

Proof Let

$$T_h \; : \; W \longrightarrow W$$

$$\omega \longmapsto \omega + \mathfrak{e}(h)$$

whose inverse is T_{-h}. Equation (1.18) can be rewritten as

$$\mathbf{E}[F \circ T_h] = \mathbf{E}[F \Lambda_h].$$

It is equivalent to

$$\mathbf{E}[F \circ T_{-h} \, \Lambda_h] = \mathbf{E}[F]. \qquad (1.19)$$

This means that the pushforward of the measure $\Lambda_h \mu$ by the map T_{-h} is the Wiener measure μ. In view of (1.13), we have to prove that for any $z \in W^*$,

$$\int_W \exp\Big(i \, \langle z, \omega - \mathfrak{e}(h) \rangle_{W^*, W} \Big) \exp\Big(\delta h(\omega) - \frac{1}{2} \|h\|_{\mathcal{H}}^2 \Big) \mathrm{d}\mu(\omega)$$

$$= \exp\Big(-\frac{1}{2} \|\mathfrak{e}^*(z)\|_{\mathcal{H}}^2 \Big). \qquad (1.20)$$

Remark that

$$i \, \langle z, \omega - \mathfrak{e}(h) \rangle_{W^*, W} + \delta h(\omega) - \frac{1}{2} \|h\|_{\mathcal{H}}^2$$

$$= i \, \langle z, \omega \rangle_{W^*, W} + \delta h(\omega) - i \, \langle z, \mathfrak{e}(h) \rangle_{W^*, W} - \frac{1}{2} \|h\|_{\mathcal{H}}^2$$

$$= i \delta\big(\mathfrak{e}^*(z) - ih \big)(\omega) - i \, \big\langle \mathfrak{e}^*(z), h \big\rangle_{\mathcal{H}} - \frac{1}{2} \|h\|_{\mathcal{H}}^2. \qquad (1.21)$$

In view of the definition of the Wiener integral,

$$\int_W \exp\Big(i \delta\big(\mathfrak{e}^*(z) - ih \big)(\omega) \Big) \mathrm{d}\mu(\omega) = \exp\Big(-\frac{1}{2} \|\mathfrak{e}^*(z) - ih\|_{\mathcal{H}}^2 \Big).$$

Since $\mathcal{H}$ is a real (not a complex) Hilbert space,

$$\|\mathfrak{e}^*(z) - ih\|_{\mathcal{H}}^2 = \|\mathfrak{e}^*(z)\|_{\mathcal{H}}^2 - \|h\|_{\mathcal{H}}^2 - 2i \, \langle \mathfrak{e}^*(z), h \rangle_{\mathcal{H}}. \qquad (1.22)$$

Plug (1.22) into (1.21) to get (1.20). $\square$

A Quick Refresher About Hilbert Spaces

We shall often encounter partial functions: For a function of several variables, say $f(t, s)$, we denote by $f(t, .)$ the partial function

$$f(t, .) : E \longrightarrow \mathbf{R}$$

$$s \longmapsto f(t, s).$$

Definition 1.12 A Hilbert space $(H, \langle ., . \rangle_H)$ is a vector space H, which is complete for the topology induced by the scalar product $\langle ., . \rangle_H$.

Recall that a metric space E is separable whenever there exists a denumerable family which is dense: There exists $(x_n, n \geq 1)$ such that for any $\epsilon > 0$, any $x \in X$, one can find some x_n such that $d(x, x_n) < \epsilon$. By construction, the set of rational numbers is such a set in $\mathbf{R}$. All the spaces we are going to consider, even the seemingly ugliest, are separable hence we can safely forget this subtlety.

Theorem 1.9 *Any separable Hilbert space H admits a complete orthonormal basis (CONB for short) $(e_n, n \geq 1)$, i.e., on the one hand*

$$\langle e_n, e_m \rangle_H = \mathbf{1}_{\{n\}}(m)$$

and on the other hand, any $x \in H$ can be written

$$x = \sum_{n=1}^{\infty} \langle x, e_n \rangle_H \, e_n$$

which means

$$\lim_{N \to \infty} \left\| x - \sum_{n=1}^{N} \langle x, e_n \rangle_H \, e_n \right\|_H = 0.$$

We will use repeatedly in diverse contexts the Parseval inequality that says the following.

Corollary 1.2 (Parseval) *Let $(e_n, n \geq 1)$ be a CONB. For any $x \in H$,*

$$\|x\|_{H^2} = \sum_{n=1}^{\infty} \langle x, e_n \rangle_H^2 \quad \text{and} \quad \langle x, y \rangle_H = \sum_{n=1}^{\infty} \langle x, e_n \rangle_H \, \langle y, e_n \rangle_H .$$

The classical example of Hilbert spaces is the space of square integrable functions from a measurable space $(E; m)$ into $\mathbf{R}$:

$$L^2(E \to \mathbf{R}; m) = \left\{ f : E \to \mathbf{R}, \int_E |f(x)|^2 \mathrm{d}m(x) < \infty \right\},$$

with the scalar product

$$\langle f, g \rangle_{L^2(E \to \mathbf{R}; m)} = \int_E f(x) g(x) dm(x).$$

Self-reproducing Hilbert Spaces

Assume we are given a symmetric function R on $E \times E$ satisfying

$$\sum_{k,l=1}^n R(t_k, t_l) c_k c_l \geq 0$$

for any $n \geq 1$, any $t_1, \cdots, t_n \in E$, and any $c_1, \cdots, c_n \in \mathbf{R}$, with equality if and only if $c_k = 0$ for all k. Then, R is said to be symmetric positive definite kernel.

Definition 1.13 Consider $H_0 = \text{span}\{R(t, .), t \in E\}$ and define an inner product on H_0 by

$$\langle R(t, .), R(s, .)\rangle_{H_0} = R(t, s). \tag{1.23}$$

Then, H is the completion of H_0 with respect to this inner product: The set of functions of the form

$$f(s) = \sum_{i=1}^{\infty} \alpha_i R(t_i, s)$$

for some denumerable family $(t_k, k \geq 1)$ of elements of E and some real numbers $(\alpha_k, k \geq 1)$ such that

$$\sum_{i=1}^{\infty} \alpha_i^2 R(t_i, t_i) < \infty.$$

Compact Maps in Hilbert Spaces

Definition 1.14 A linear map T between two Hilbert spaces H_1 and H_2 is said to be compact whenever the image of any bounded subset in H_1 is a relatively compact subset (i.e., its closure is compact) in H_2. It can be written: For any $h \in H_1$

$$Th = \sum_{n=1}^{\infty} \lambda_n \langle f_n, h \rangle_{H_1} g_n$$

where $(f_n, n \geq 1)$ and $(g_n, n \geq 1)$ are orthonormal sets of respectively H_1 and H_2. Moreover, $(\lambda_n, n \geq 1)$ is a sequence of positive real numbers with sole accumulation point zero. If for some rank N, $\lambda_n = 0$ for $n \geq N$, the operator is said to be of finite rank.

Among those operators, some will play a crucial rôle in the sequel.

Definition 1.15 (Trace Class Operators) Let H be a Hilbert space and $(e_n, n \geq 1)$ be a CONB on H. A linear map A from H into itself is said to be trace-class whenever

$$\sum_{n \geq 1} |\langle Ae_n, e_n \rangle| < \infty.$$

Then, its trace is defined as

$$\text{trace}(A) = \sum_{n \geq 1} \langle Ae_n, e_n \rangle.$$

In the decomposition of Definition 1.14, this means that

$$\sum_{n=1}^{\infty} |\lambda_n| < \infty.$$

Definition 1.16 (Hilbert–Schmidt Operators) Let H_1 and H_2 be two Hilbert space and $(e_n, n \geq 1)$ (resp. $(f_p, p \geq 1)$) a CONB of H_1 (resp. H_2). A linear map A from H_1 into H_2 is said to be Hilbert–Schmidt whenever

$$\|A\|_{\text{HS}}^2 = \sum_{n \geq 1} \|Ae_n\|_{H_2}^2 = \sum_{n \geq 1} \sum_{p \geq 1} \langle Ae_n, f_p \rangle_{H_2}^2 < \infty.$$

If $H_1 = H_2$, in the decomposition of Definition 1.14, this means that

$$\sum_{n=1}^{\infty} \lambda_n^2 < \infty.$$

Note that a linear map from H into itself can be described by an infinite matrix: To characterize A, since H has a basis, it is sufficient to determine its values on this basis. This means that A is completely determined by the family

$(\langle Ae_n, e_k\rangle_H$, $n, k \geq 1)$, which is nothing but a kind of an infinite matrix. We can also write

$$\langle Ae_n, e_k\rangle_H = \langle A, e_n \otimes e_k\rangle_{H\otimes H},$$

so that A appears as a linear map on $H \otimes H$.

Theorem 1.10 *If $H = L^2(E \to \mathbf{R}; \mu)$ and A is Hilbert–Schmidt, then there exists a kernel that we still denote by $A : H \times H \to \mathbf{R}$ such that for any $f \in H$,*

$$Af(x) = \int_E A(x, y) f(y)\mathrm{d}\mu(y)$$

and

$$\|A\|_{\mathrm{HS}}^2 = \iint_{E\times E} |A(x, y)|^2\mathrm{d}\mu(x)\mathrm{d}\mu(y).$$

Theorem 1.11 (Composition of Hilbert–Schmidt Maps) *With the same notations as above, the composition of two Hilbert–Schmidt is trace class.*

Actually, this is an equivalence: A trace-class map can always be written as the composition of two Hilbert–Schmidt operators. Moreover,

$$|\operatorname{trace}(A \circ B)| \leq \sum_{n\geq 1}|\langle A \circ Be_n, e_n\rangle_H| \leq \|A\|_{\mathrm{HS}}\|B\|_{\mathrm{HS}}. \tag{1.24}$$

Lemma 1.3 (Composition of Integral Maps) *If $H = L^2(E \to \mathbf{R}; \mu)$ and A, B are Hilbert–Schmidt maps on H. Then, $B \circ A$ is trace-class and*

$$\operatorname{trace}(B \circ A) = \iint_{E\times E} B(x, y)A(y, x)\mathrm{d}\mu(x)\mathrm{d}\mu(y).$$

Proof We must verify the finiteness of

$$\sum_{n\geq 1} |\langle BAe_n, e_n\rangle_H|.$$

By the definition of the adjoint, applying twice the Cauchy–Schwarz inequality, we have

$$\sum_{n\geq 1} |\langle BAe_n, e_n\rangle_H| = \sum_{n\geq 1}|\langle Ae_n, B^*e_n\rangle_H| \leq \sum_{n\geq 1} \|Ae_n\|_H\|B^*e_n\|_H$$

$$\leq \left(\sum_{n\geq 1} \|Ae_n\|_H^2\right)^{1/2} \left(\sum_{n\geq 1} \|B^*e_n\|_H^2\right)^{1/2} = \|A\|_{\mathrm{HS}}\|B^*\|_{\mathrm{HS}}.$$

The Parseval identity (twice) yields

$$\text{trace}(B \circ A) = \sum_{n\geq 1} \langle Ae_n, B^*e_n \rangle_H = \sum_{n\geq 1}\sum_{k\geq 1} \langle Ae_n, e_k \rangle_H \langle B^*e_n, e_k \rangle_H$$

$$= \sum_{n\geq 1}\sum_{k\geq 1} \langle A, e_k \otimes e_n \rangle_{H\otimes H} \langle B^*, e_k \otimes e_n \rangle_{H\otimes H} = \langle A, B^* \rangle_{H\otimes H} .$$

By the identification of A, B and their kernel,

$$\langle A, B^* \rangle_{H\otimes H} = \iint_{H\times H} A(x,y)B^*(x,y)\mathrm{d}\mu(x)\mathrm{d}\mu(y)$$

$$= \iint_{H\times H} A(x,y)B(y,x)\mathrm{d}\mu(x)\mathrm{d}\mu(y).$$

The proof is thus complete.   □

Example (Hilbert–Schmidt Embeddings in Fractional Liouville Spaces) Since $I^\alpha \circ I^\beta = I^{\alpha+\beta}$, we have

$$I_{\beta,2} \subset I_{\alpha,2} \text{ for any } \beta > \alpha.$$

Lemma 1.4 *The embedding* $\mathfrak{e}$ *of* $I_{\beta,2}$ *into* $I_{\alpha,2}$ *is Hilbert–Schmidt if and only if* $\beta - \alpha > 1/2$.

Proof Let $(e_n, n \geq 1)$ be CONB of $L^2([0,1])$ and set $h_n = I^\beta e_n$. Then $(h_n, n \geq 1)$ is a CONB of $I_{\beta,2}$. We must prove that

$$\sum_{n=1}^{\infty} \|\mathfrak{e}(h_n)\|_{I_{\alpha,2}}^2 < \infty.$$

By the very definition of the norm in $I_{\alpha,2}$, this is equivalent to show

$$\sum_{n=1}^{\infty} \|I^{\beta-\alpha}(e_n)\|_{L^2}^2 < \infty.$$

But this latter sum turns to be equal to the Hilbert–Schmidt norm of $I^{\beta-\alpha}$ viewed as a linear map from L^2 into itself. In view of Proposition 1.10, $I^{\beta-\alpha}$ is Hilbert–Schmidt if and only if

$$\iint_{[0,1]^2} |t-s|^{2((\beta-\alpha)-1)}\mathrm{d}s\mathrm{d}t < \infty.$$

This only happens if $\beta - \alpha > 1/2$.   □

1.4   Problems

1.1 (Dual of $L^2([0, 1] \to \mathbf{R}; \ell)$) Since we identified $I^{1,2}$ and its dual, we cannot identify $L^2([0, 1] \to \mathbf{R}; \ell)$ and its dual as usual. Show that the dual of $L^2([0, 1] \to \mathbf{R}; \ell)$ can be identified to $I_-^1(I^{1,2})$ where

$$I_-^1 f(t) = \int_t^1 f(s) \mathrm{d}s.$$

1.2 (Brownian Measure on $I_{\alpha,2}$) From Theorem 1.4, we know that $I_{\alpha,2} \subseteq L^2$ for any $\alpha > 0$.

1. Show that this embedding is Hilbert–Schmidt if and only if $\alpha > 1/2$.

   For $\alpha > 1/2$, $I_{\alpha,2} \subseteq \mathrm{Hol}(\alpha - 1/2) \subset C$ hence, the Dirac measure ϵ_τ belongs to $I_{\alpha,2}^*$. Let j_α be the canonical isometry between $I_{\alpha,2}^*$ and $I_{\alpha,2}$.

2. Show that

$$j_\alpha(\epsilon_\tau) = \frac{1}{\Gamma(\alpha)} I^\alpha \left((\tau - .)^{\alpha-1} \right).$$

3. Following the proof of Theorem 1.5, show that $(S_n, \, n \geq 0)$ as defined in (1.7) is convergent in $I_{\alpha,2}$ for $\alpha < 1/2$.

   It is important to remark that $(\dot{h}_n, \, n \geq 0)$ *is an orthonormal family of* $L^2([0, 1] \to \mathbf{R}; \ell)$.

4. Show that for any $z \in I_{\alpha,2}$,

$$\mathbf{E}\left[e^{i \langle z, \sum_n X_n h_n \rangle_{I_{\alpha,2}}} \right] = \exp(-\frac{1}{2} \langle V_\alpha z, z \rangle_{I_{\alpha,2}})$$

   where

$$V_\alpha = I^\alpha \circ I^{1-\alpha} \circ (I^{1-\alpha})^* \circ (I^\alpha)^{-1}.$$

1.3 (Wiener Space of the Brownian Bridge) The Brownian bridge W is the centered Gaussian process whose covariance kernel is given by

$$\mathbf{E}[W(t)W(s)] = s \wedge t \, (1 - s \vee t).$$

Alternatively, it can be described by a transformation of the Brownian motion:

$$W(t) \stackrel{\text{dist.}}{=} B(t) - t B(1),$$

where B is an ordinary Brownian motion.

Let

$$P : W \longrightarrow W$$

$$f \longmapsto \left(t \mapsto f(t) - tf(1)\right).$$

Let W_0 be the elements of W, which are null at time 1 and $\mathcal{H}_0 = W_0 \cap \mathcal{H}$.

1. Show that P is an orthogonal projection from $\mathcal{H}$ to $\mathcal{H}_0$. Prove that

$$\mathcal{H} = \mathcal{H}_0 \oplus \operatorname{span} h_0$$

   where $h_0(t) = t$ as in the definition of the basis of $\mathcal{H}$, see (1.6).
2. Derive that $(h_n, \ n \geq 1)$ is a complete orthonormal basis of $\mathcal{H}_0$.
3. Show that for $h \in \mathcal{H}_0$, the law of $W + h$ is absolutely continuous with respect to the distribution of W.
4. Let δ_W be the Wiener integral with respect to W. Show that

$$\delta_W(s \wedge . - s * .) = W(s).$$

5. Alternatively, show that

$$\sum_{n \geq 1} X_n h_n,$$

   where $(X_n, \ n \geq 1)$ is a family of independent standard Gaussian variables and $(h_n, \ n \geq 1)$ the basis mentioned in (1.6), converges with probability 1, in $\mathcal{H}_0$, to a Gaussian process that has the distribution of W.

1.5   Notes and Comments

The construction of the Wiener measure dates back to the Donsker's Theorem [2], improved ten years later by Lamperti [6]. A more abstract version of the construction of an abstract Wiener space is to consider a triple made of a Hilbert space $\mathcal{H}$, a Banach space W, and a continuous injective map ϵ from $\mathcal{H}$ into W, with dense image and which is *radonifying* (meaning that it transforms a cylindric measure into a true Radon measure). If W is an Hilbert space, this amounts to assume that ϵ is Hilbert–Schmidt (see [1]). Radonifying functions are the subject of the monography [9]. Proposition 25.6.3 of [7] states that the canonical embedding of $\mathcal{H}$ into any $W_{\alpha,p}$ is indeed radonifying.

The presentation given here is inspired by Stroock [11] and Itô and Nisio [5]. Another construction can be found in [10]. For details on Hilbert spaces and operators on such spaces, the reader could consult any book relative to functional analysis like [12] or [3, 4] for the not faint of heart.

The properties of fractional integrals that will be needed essentially in the chapter about fractional Brownian motion (see Chap. 4) can be found in the Bible for almost everything about fractional calculus [8].

References

1. L. Coutin, L. Decreusefond, Stein's method for Brownian approximations. Commun. Stoch. Anal. **7**(3), 349–372 (2013)
2. M.D. Donsker, An invariance principle for certain probability limit theorems. Mem. Am. Math. Soc. **6**, 00427, 12 (1951)
3. N. Dunford, J.T. Schwartz, *Linear Operators. Part I* (Wiley Classics Library, New York, 1988)
4. N N. Dunford, J.T Schwartz, *Linear Operators. Part II* (Wiley Classics Library, New York, 1988)
5. K. Itô, M. Nisio, On the convergence of sums of independent Banach space valued random variables. Osaka J. Math. **5**, 35–48 (1968)
6. J. Lamperti, On convergence of stochastic processes. Trans. Am. Math. Soc. **104**, 430–435 (1962)
7. A. Pietsch, *Operator Ideals*, vol. 20 (North-Holland, Amsterdam, 1980)
8. S.G. Samko, A.A. Kilbas, O.I. Marichev, *Fractional Integrals and Derivatives* (Gordon and Breach Science, Philadelphia, 1993)
9. L. Schwartz, *Séminaire Laurent Schwartz 1969–1970: Applications Radonifiantes* (Centre de Mathématiques, Paris, 1970)
10. H.-H. Shih, On Stein's method for infinite-dimensional gaussian approximation in abstract Wiener spaces. J. Funct. Anal. **261**(5), 1236–1283 (2011)
11. D.W. Stroock, Abstract Wiener space, revisited. Commun. Stoch. Anal. **2**(1), 145–151 (2008)
12. K. Yosida, *Functional Analysis* (Springer, Berlin, 1995)

Chapter 2
Gradient and Divergence

Abstract If we put a measure on a Banach space, functions defined on it become random variables and are thus defined up to a negligeable set. This ruins the possibility to use Fréchet calculus on such a space. The Cameron-Martin theorem says that we must restrict the directions in which we can derive to a dense but negligeable set. Hence the importance of the Gross-Sobolev-Malliavin gradient we define now.

2.1 Gradient

If our objective is to define a differential calculus on the Banach space W, why don't we use the notion of Fréchet derivative? A function $F : W \to \mathbf{R}$ is said to be Fréchet differentiable if there exists a continuous linear operator $A : W \to \mathbf{R}$ such that

$$\lim_{\epsilon \to 0} \epsilon^{-1} \left\| F(\omega + \epsilon\omega') - F(\omega) - \epsilon A(\omega') \right\|_W = 0 \qquad (2.1)$$

for any $\omega \in W$ and any $\omega' \in W$. In particular, a Fréchet differentiable function is continuous. One of the most immediate function we can think of is the so-called Itô map which sends a sample-path ω to the corresponding sample-path of the solution of a well defined stochastic differential equation. It is well known (see [3, Section 3.3] for instance) that in dimension higher than one, this map is not continuous. This induces that the notion of Fréchet derivative is not well suited to a differential calculus on the Wiener space. Moreover, since we work on a probability space, measurable functions F from W into $\mathbf{R}$ are random variables, meaning that they are defined up to a negligible set. To avoid any inconsistency in a formula like (2.1), we must ensure that

$$(F = G \ \mu \text{ a.s.}) \implies (F(. + \omega') = G(. + \omega') \ \mu \text{ a.s.})$$

© The Author(s), under exclusive license to Springer Nature Switzerland AG 2022

L. Decreusefond, *Selected Topics in Malliavin Calculus*,
Bocconi & Springer Series 10, https://doi.org/10.1007/978-3-031-01311-9_2

for any ω'. With the notations of Theorem 1.8, this requires that $T_{\omega'}^{\#}\mu$ (the push-forward of the measure μ by the translation map $T_{\omega'}$) to be absolutely continuous with respect to the Wiener measure μ. This fact is granted only if ω' belongs to $I_{1,2}$. These two reasons mean that we are to define the directional derivative of F in a restricted class of possible perturbations.

The basic definition of the differential of a function $f : \mathbf{R}^n \to \mathbf{R}$ is to consider the limit of

$$\lim_{\varepsilon \to 0} \varepsilon^{-1}(f(x + \varepsilon y)) - f(x)).$$

For modern applications, this definition is insufficient as it says nothing on the integrability of the so-defined derivative. This is where the notion of Sobolev spaces takes its paramount importance. One of the possible definition of the Sobolev space $H^{1,2}(\mathbf{R}^n)$ is to define it as the completion of $C_c^1(\mathbf{R}^n; \mathbf{R})$, the space of C^1 class functions with compact support, with respect to the norm

$$\|f\|_{H^{1,2}(\mathbf{R}^n)} = \left(\|f\|_{L^2\left(\mathbf{R}^n \to \mathbf{R};\, \ell\right)}^2 + \sum_{j=1}^{n} \|\partial_j f\|_{L^2\left(\mathbf{R}^n \to \mathbf{R};\, \ell\right)}^2 \right)^{1/2}.$$

We more or less copy this approach here, replacing the space $C_c^1(\mathbf{R}^n; \mathbf{R})$ by the space of cylindrical functionals and then defining the gradient only in the directions allowed by the Cameron–Martin space. To pursue the reasoning, we need to prove that the so-defined gradient is closable, i.e., if we choose different sequences approaching the same functions in some $L^p(\mathbf{W} \to \mathbf{R}; \mu)$, the limits of their gradient should be the same. This turns to be guaranteed by (a consequence of) the quasi-invariance formula (1.18).

Recall the diagram

$$
\begin{array}{ccc}
\mathbf{W}^* & \xrightarrow{\ e^* \ } & \mathcal{H}^* = (I_{1,2})^* \\
 & & \| \\
L^2 & \xrightarrow{\ I^1 \ } \mathcal{H} = I_{1,2} & \xrightarrow{\ e \ } \mathbf{W}
\end{array}
$$

and that μ is the Wiener measure on $\mathbf{W}$. We first recall the definition of the Schwartz space on $\mathbf{R}^n$.

Definition 2.1 The Schwartz space on $\mathbf{R}^n$, denoted by Schwartz($\mathbf{R}^n$), is the set of C^∞ functions from $\mathbf{R}^n$ to $\mathbf{R}$ whose all derivatives are rapidly decreasing: f belongs to Schwartz($\mathbf{R}^n$) if for any $\alpha = (\alpha_1, \cdots, \alpha_n) \in \mathbf{N}^n$ and any $\beta = (\beta_1, \cdots, \beta_n) \in (\mathbf{R}^+)^n$,

$$\sup_{x \in \mathbf{R}^n} \left| x^\beta \partial^\alpha f(x) \right| < \infty.$$

Definition 2.2 A function $F : W \to \mathbf{R}$ is said to be cylindrical if there exist an integer n, a function $f \in \text{Schwartz}(\mathbf{R}^n)$, $(h_1, \cdots, h_n) \in \mathcal{H}^n$ such that

$$F(\omega) = f\big(\delta h_1(\omega), \cdots, \delta h_n(\omega)\big).$$

The set of such functionals is denoted by $\mathcal{S}$.

Theorem 2.1 *The set $\mathcal{S}$ is dense in $L^p(W \to \mathbf{R}; \mu)$.*

Proof Let $\mathcal{D}_n$ be the dyadic subdivision of mesh 2^{-n} of $[0, 1]$ and $\mathcal{F}_n = \sigma\{B(t), t \in \mathcal{D}_n\}$. Any continuous function can be approximated by its affine interpolation on the dyadic subdivisions hence $\vee_n \mathcal{F}_n = \mathcal{F}$ and the $L^p(W \to \mathbf{R}; \mu)$ convergence theorem for martingales says that

$$\mathbf{E}[F \mid \mathcal{F}_n] \xrightarrow[L^p(W \to \mathbf{R}; \mu)]{n \to \infty} F.$$

For $\epsilon > 0$, let n such that $\| F - \mathbf{E}[F \mid \mathcal{F}_n] \|_{L^p(W \to \mathbf{R}; \mu)} < \epsilon$. The Doob Lemma entails that there exists ψ_n measurable from $\mathbf{R}^{2^n}$ to $\mathbf{R}$ such that

$$\mathbf{E}[F \mid \mathcal{F}_n] = \psi_n(B(t), t \in \mathcal{D}_n).$$

Let μ_n be the distribution of the Gaussian vector $(B(t), t \in \mathcal{D}_n)$,

$$\int |\psi_n|^p \mathrm{d}\mu_n = \mathbf{E}\big[|\mathbf{E}[F \mid \mathcal{F}_n]|^p\big] \leq \mathbf{E}\big[|F|^p\big] < \infty.$$

That means that ψ_n belongs to $L^p(\mathbf{R}^n \to \mathbf{R}; \mu_n)$ hence for any $\epsilon > 0$, there exists $\varphi_\epsilon \in \mathcal{S}(\mathbf{R}^{2^n})$ such that $\| \psi_n - \varphi_\epsilon \|_{L^p(\mathbf{R}^n \to \mathbf{R}; \mu_n)} < \epsilon$. Then, $\varphi_\epsilon(B(t), t \in \mathcal{D}_n)$ belongs to $\mathcal{S}$ and is within distance 2ϵ of F in $L^p(W \to \mathbf{R}; \mu)$. $\square$

The gradient is first defined on cylindrical functionals.

Definition 2.3 Let $F \in \mathcal{S}, h \in \mathcal{H}$, with $F = f(\delta h_1, \cdots, \delta h_n)$. Set

$$\nabla F = \sum_{j=1}^n (\partial_j f)\big(\delta h_1, \cdots, \delta h_n\big) h_j,$$

so that

$$\langle \nabla F, h \rangle_{\mathcal{H}} = \sum_{j=1}^n (\partial_j f)\big(\delta h_1, \cdots, \delta h_n\big) \langle h_j, h \rangle_{\mathcal{H}}.$$

This definition is coherent with the natural definition of directional derivative.

Lemma 2.1 *For $F \in S$, for $h \in \mathcal{H}$, we have*

$$\langle \nabla F(\omega), h \rangle_{\mathcal{H}} = \lim_{\epsilon \to 0} \frac{1}{\epsilon} \Big(F(\omega + \epsilon h) - F(\omega) \Big).$$

Proof For $F(\omega) = f\Big(\delta h_1(\omega), \cdots, \delta h_n(\omega) \Big)$,

$$F(\omega + \epsilon h) = f\Big(\delta h_1(\omega + \epsilon h), \cdots, \delta h_n(\omega + \epsilon h) \Big)$$
$$= f\Big(\delta h_1(\omega) + \epsilon \langle h_1, h \rangle_{\mathcal{H}}, \cdots, \delta h_n(\omega) + \epsilon \langle h_n, h \rangle_{\mathcal{H}} \Big)$$

because of (1.15). Now then, we apply the classical chain rule to derive with respect to ϵ and substitute 0 to ϵ to obtain

$$\frac{d}{d\epsilon} F(\omega + \epsilon h) \Big|_{\epsilon=0} = \sum_{j=1}^{n} (\partial_j f)\Big(\delta h_1(\omega), \cdots, \delta h_n(\omega) \Big) \langle h_j, h \rangle_{\mathcal{H}}.$$

The proof is thus complete.   □

In view of Lemma 2.1, we see that S is an algebra for the ordinary product.

Corollary 2.1 *For $F \in S$, $\phi \in$ Schwartz$(\mathbf{R})$,*

$$\nabla (FG) = F \nabla G + G \nabla F \tag{2.2}$$
$$\nabla \phi(F) = \phi'(F) \nabla F. \tag{2.3}$$

Example (Derivative of $f(B(t))$) Recall that $B(t) = \delta(t \wedge .)$. Hence, for $f \in$ Schwartz$(\mathbf{R})$,

$$\nabla f(B(t)) = f'(B(t)) \nabla (\delta(t \wedge .)) = f'(B(t)) t \wedge .$$

As shows the last example, the previous definition entails that each ω, $\nabla F(\omega)$ is an element of $\mathcal{H}$, i.e., a differentiable function whose derivative is square integrable. Hence, we can speak of $(\omega, s) \longmapsto \nabla_s F(\omega)$. This means that ∇F can be viewed as an $\mathcal{H}$-valued random variable or as a process with differentiable paths. In the setting of Malliavin calculus, we adopt the former point of view. As such it is now natural to discuss the integrability of the random variable ∇F.

Before going further it may be worth looking below for some elements about tensor products of Banach spaces.

Theorem 2.2 *For $F \in S$, ∇F belongs to $L^p(\mathrm{W} \to \mathcal{H}; \mu)$ for any $p \geq 1$.*

Proof **Step 1** Assume $p > 1$. Since

$$L^p(\mathrm{W} \to \mathcal{H}; \mu) \simeq L^p(\mathrm{W} \to \mathbf{R}; \mu) \otimes \mathcal{H},$$

we have

$$\left(L^p\left(W \to \mathcal{H};\ \mu\right)\right)^* \simeq L^q\left(W \to \mathbf{R};\ \mu\right) \otimes \mathcal{H}$$

where $q = p/(p-1)$.

Step 2 Consider the set

$$B_{q,\mathcal{H}} = \{(k,\ G) \in \mathcal{H} \times L^q\left(W \to \mathbf{R};\ \mu\right),\ \|k\|_{\mathcal{H}} = 1,\ \|G\|_{L^q\left(W \to \mathbf{R};\ \mu\right)} = 1\}.$$

Let $F = f(\delta h)$, for $p > 1$, the Proposition 2.1 says that to prove that the p-norm of ∇F is finite, it is sufficient to show that

$$\sup_{(k,G) \in B_{q,\mathcal{H}}} \left|\langle \nabla F, k \otimes G \rangle_{L^p\left(W \to \mathcal{H};\ \mu\right),\ L^q\left(W \to \mathcal{H};\ \mu\right)}\right| < \infty.$$

Recall that $L^p\left(W \to \mathcal{H};\ \mu\right) \simeq L^p\left(W \to \mathbf{R};\ \mu\right) \otimes \mathcal{H}$. Thus, for $T \in L^p\left(W \to \mathbf{R};\ \mu\right)$ and $l \in \mathcal{H}$, by the very definition of the duality bracket (see (2.39)),

$$\langle T \otimes l, G \otimes k \rangle_{L^p\left(W \to \mathcal{H};\ \mu\right),\ L^q\left(W \to \mathcal{H};\ \mu\right)} = \langle T, G \rangle_{L^p\left(W \to \mathbf{R};\ \mu\right),\ L^q\left(W \to \mathbf{R};\ \mu\right)} \langle l, k \rangle_{\mathcal{H}}$$

$$= \mathbf{E}[FG]\langle l, k \rangle_{\mathcal{H}}$$

$$= \mathbf{E}\left[\langle F \otimes l, G \otimes k \rangle_{\mathcal{H}}\right].$$

By density of the pure tensor products, we get

$$\left|\langle \nabla F, k \otimes G \rangle_{L^p\left(W \to \mathcal{H};\ \mu\right),\ L^q\left(W \to \mathcal{H};\ \mu\right)}\right| = \left|\mathbf{E}\left[\langle \nabla F, k \rangle_{\mathcal{H}} G\right]\right|$$

$$= \left|\mathbf{E}\left[f'(\delta h)G\right]\langle k, h \rangle_{\mathcal{H}}\right|$$

$$\leq \|f'\|_\infty \|G\|_{L^q\left(W \to \mathbf{R};\ \mu\right)} \|h\|_{\mathcal{H}} \|k\|_{\mathcal{H}}.$$

Hence the supremum over $B_{q,\mathcal{H}}$ is finite. The same proof can be applied when $F = f(\delta h_j,\ 1 \leq j \leq m)$.

Step 3 For $p = 1$, the previous considerations no longer prevail since an L^1 space is not reflexive so that we cannot apply (2.41). However, it is sufficient to see that $L^p\left(W \to \mathcal{H};\ \mu\right)$ is included in $L^1\left(W \to \mathcal{H};\ \mu\right)$.

$\square$

It is an exercise left to the reader to see that the map

$$\mathrm{Id} \otimes I^{-1} : L^p\left(W \to \mathbf{R};\ \mu\right) \otimes \mathcal{H} \longrightarrow L^p\left(W \to \mathbf{R};\ \mu\right) \otimes L^2\left([0,1] \to \mathbf{R};\ \ell\right)$$

$$F \otimes h \longmapsto F \otimes \dot{h}$$

is continuous. Moreover, Theorem 2.2 means for any $F \in \mathcal{S}$, ∇F belongs to $L^p(W \to \mathbf{R}; \mu) \otimes \mathcal{H}$. Hence there exists an element $\dot{\nabla} F$ of $L^p(W \to \mathbf{R}; \mu) \otimes L^2([0, 1] \to \mathbf{R}; \ell)$ such that

$$\langle \nabla F, h \rangle_{\mathcal{H}} = \int_0^1 \dot{\nabla}_s F \, \dot{h}(s) \mathrm{d}s$$

$$\text{and } \|F\|_{L^p(W \to \mathcal{H}; \mu)} = \mathbf{E}\left[\left(\int_0^1 |\dot{\nabla}_s F|^2 \mathrm{d}s\right)^{p/2}\right]^{1/p}.$$

We now have a nice Banach space into which our gradient lives. The idea is then to extend it by density, i.e., take a sequence $(F_n, n \geq 1)$ of cylindrical functions that converges in $L^p(W \to \mathbf{R}; \mu)$ to a function F and say that if the sequence of gradients $(\nabla F_n, n \geq 1)$ converge to something in $L^p(W \to \mathcal{H}; \mu)$, then F is differentiable and its gradient is the latter limit. For this procedure to be valid, we need to ensure that the limit does not depend on the approximating sequence. This is the rôle of the notion of closability.

Theorem 2.3 ∇ *is closable in* $L^p(W \to \mathcal{H}; \mu)$ *for* $p > 1$.

This means that if $F_n \in \mathcal{S}$ *tends to 0 in* $L^p(W \to \mathbf{R}; \mu)$ *and* ∇F_n *tends to* η *in* $L^p(W \to \mathcal{H}; \mu)$, *then* $\eta = 0$.

Integration by Parts Is a Consequence of Invariance
The classical integration by parts formula reads as

$$\int_{\mathbf{R}} f(x)g'(x)\mathrm{d}x = -\int_{\mathbf{R}} f'(x)g(x)\mathrm{d}x \qquad (2.4)$$

if f and g do vanish at infinity. It can be seen as a consequence of the invariance of the Lebesgue measure with respect to translations. Actually, we have for any $y \in \mathbf{R}$,

$$\int_{\mathbf{R}} f(x + y)g(x + y)\mathrm{d}x = \int_{\mathbf{R}} f(x)g(x)\mathrm{d}x.$$

The right-hand side does not depend on y, hence if we differentiate the left-hand side with respect to y at $y = 0$, we obtain (2.4).

The Wiener measure is not invariant but only quasi-invariant; this gives an additional term in the integration by parts formula.

Lemma 2.2 (Integration by Parts) *For* F *and* G *cylindrical, for* $h \in \mathcal{H}$,

$$\mathbf{E}[G \langle \nabla F, h \rangle_{\mathcal{H}}] = -\mathbf{E}[F \langle \nabla G, h \rangle_{\mathcal{H}}] + \mathbf{E}[FG\, \delta h]. \qquad (2.5)$$

Proof The Cameron–Martin theorem says that

$$\int_W F(\omega+\epsilon h)G(\omega+\epsilon h)d\mu(\omega) = \int_W F(\omega)G(\omega)\exp\left(\epsilon\delta h(\omega) - \frac{\epsilon^2}{2}\|h\|_{\mathcal{H}}^2\right)d\mu(\omega).$$

Differentiate both sides with respect to ϵ, at $\epsilon = 0$, to obtain

$$\mathbf{E}\left[F\langle\nabla G, h\rangle_{\mathcal{H}}\right] + \mathbf{E}\left[G\langle\nabla F, h\rangle_{\mathcal{H}}\right] = \mathbf{E}[FG\,\delta h],$$

which corresponds to Eq. (2.5). □

Proof of Theorem 2.3 Let $(F_n, n \geq 1)$, which tends to 0 in $L^p(W \to \mathbf{R}; \mu)$ and such that ∇F_n tends to η in $L^p(W \to \mathcal{H}; \mu)$. Then the right-hand side of Eq. (2.5) tends to 0. On the other hand, by definition of the convergence in $L^p(W \to \mathcal{H}; \mu)$,

$$\mathbf{E}\left[G\langle\nabla F_n, h\rangle_{\mathcal{H}}\right] \xrightarrow{n\to\infty} \langle\eta, h\otimes G\rangle_{L^p(W\to\mathcal{H};\mu),\,L^q(W\to\mathcal{H};\mu)}.$$

It means that for any $h \in \mathcal{H}$ and $G \in \mathcal{S}$,

$$\langle\eta, h\otimes G\rangle_{L^p(W\to\mathcal{H};\mu),\,L^q(W\to\mathcal{H};\mu)} = 0. \tag{2.6}$$

By density of $\mathcal{S}$ in $L^p(W \to \mathbf{R}; \mu)$, (2.6) holds for $G \in L^p(W \to \mathbf{R}; \mu)$. According to Theorem 2.10, $\langle\eta, \zeta\rangle_{L^p(W\to\mathcal{H};\mu),L^q(W\to\mathcal{H};\mu)} = 0$ for any $\zeta \in L^q(W \to \mathcal{H}; \mu)$, hence $\eta = 0$. □

Definition 2.4 A functional F belongs to $\mathbb{D}_{p,1}$ if there exists $(F_n, n \geq 0)$ that converges to F in $L^p(W \to \mathbf{R}; \mu)$, such that $(\nabla F_n, n \geq 0)$ is Cauchy in $L^p(W \to \mathcal{H}; \mu)$. Then, ∇F is defined as the limit of this sequence. We put on $\mathbb{D}_{p,1}$ the norm

$$\|F\|_{p,1} = \mathbf{E}\left[|F|^p\right]^{1/p} + \mathbf{E}\left[\|\nabla F\|_{\mathcal{H}}^p\right]^{1/p}. \tag{2.7}$$

With this definition, it is not easy to determine whether a given function belongs to $\mathbb{D}_{p,1}$. The next lemma is one efficient criterion.

Lemma 2.3 *Let $p > 1$. Assume that there exists $(F_n, n \geq 0)$, which converges in $L^p(W \to \mathbf{R}; \mu)$ to F such that $\sup_n \|\nabla F_n\|_{L^p(W\to\mathcal{H};\mu)}$ is finite. Then, $F \in \mathbb{D}_{p,1}$.*

See below for the three necessary theorems of functional analysis.

Proof of Lemma 2.3 Since $\sup_n \|\nabla F_n\|_{L^p(W\to\mathcal{H};\mu)}$ is finite, there exists a sub-sequence (see Proposition 2.2), which we still denote by $(\nabla F_n, n \geq 0)$ weakly convergent in $L^p(W \to \mathcal{H}; \mu)$ to some limit denoted by η. For $k > 0$, let n_k be such that $\|F_m - F\|_{L^p(W\to\mathbf{R};\mu)} < 1/k$ for $m \geq n_k$. The Mazur's Theorem 2.3 implies that there exists a convex combination of elements of $(\nabla F_m, m \geq n_k)$ such that

$$\| \sum_{i=1}^{M_k} \alpha_i^k \nabla F_{m_i} - \eta \|_{L^p\left(W \to \mathcal{H}; \mu\right)} < 1/k.$$

Moreover, since the α_i^k are positive and sums to 1,

$$\| \sum_{i=1}^{M_k} \alpha_i^k F_{m_i} - F \|_{L^p\left(W \to \mathbf{R}; \mu\right)} = \| \sum_{i=1}^{M_k} \alpha_i^k (F_{m_i} - F) \|_{L^p\left(W \to \mathbf{R}; \mu\right)}$$

$$\leq \sum_{i=1}^{M_k} \alpha_i^k \| F_{m_i} - F \|_{L^p\left(W \to \mathbf{R}; \mu\right)} \leq \frac{1}{k}.$$

We have thus constructed a sequence

$$F^k = \sum_{i=1}^{M_k} \alpha_i^k F_{m_i}$$

such that F^k tends to F in $L^p\left(W \to \mathbf{R}; \mu\right)$ and ∇F^k converges in $L^p\left(W \to \mathcal{H}; \mu\right)$ to a limit. By the construction of $\mathbb{D}_{p,1}$, this means that F belongs to $\mathbb{D}_{p,1}$ and that $\nabla F = \eta$.   □

Example (Derivative of Doléans-Dade Exponentials) For $h \in \mathcal{H}$, the random variable

$$F = \exp\left(\delta h - \frac{1}{2}\|h\|_{\mathcal{H}}^2\right)$$

is called the Doléans-Dade exponential associated to h. The random variable F belongs to $\mathbb{D}_{p,1}$ for any $p \geq 1$ and we have

$$\nabla F = F h. \tag{2.8}$$

Remark that exp does not belong to Schwartz($\mathbf{R}$) hence we cannot apply (2.3) as is. Let

$$\exp_M : x \longmapsto \frac{M}{\sqrt{2\pi}} \int_{\mathbf{R}} \exp(y \wedge M) e^{-M^2(x-y)^2/2} dy.$$

By the properties of convolution products, $\exp_M$ belongs to Schwartz($\mathbf{R}$) and converges to exp as M goes to infinity. Moreover, in view of (2.3), we have

$$\nabla \exp_M \left(\delta h - \frac{1}{2}\|h\|_{\mathcal{H}}^2\right) = \exp_M' \left(\delta h - \frac{1}{2}\|h\|_{\mathcal{H}}^2\right) h.$$

It turns out that

$$
\exp'_M(x) = \frac{M}{\sqrt{2\pi}} \int_{\mathbf{R}} e^{x-y} \mathbf{1}_{\{x-y \leq M\}} e^{-M^2 y^2/2} dy \tag{2.9}
$$

$$
= \frac{M}{\sqrt{2\pi}} \int_{\mathbf{R}} e^{y} \mathbf{1}_{\{y \leq M\}} e^{-M^2 (x-y)^2/2} dy
$$

$$
\xrightarrow{M \to \infty} e^{x}. \tag{2.10}
$$

It thus remains to prove that

$$
\sup_M \mathbf{E}\left[\left| \exp'_M \left(\delta h - \frac{1}{2}\|h\|_{\mathcal{H}}^2 \right) \right|^p \right] < \infty. \tag{2.11}
$$

From (2.9) and Jensen inequality,

$$
\mathbf{E}\left[\left| \exp'_M \left(\delta h - \frac{1}{2}\|h\|_{\mathcal{H}}^2 \right) \right|^p \right] \leq \frac{M}{\sqrt{2\pi}} \int_{\mathbf{R}} \mathbf{E}\left[e^{p(\delta h - \frac{1}{2}\|h\|_{\mathcal{H}}^2)} \right] e^{-py} e^{-M^2 y^2/2} dy.
$$

Now then,

$$
\mathbf{E}\left[\Lambda_h^p \right] = \mathbf{E}\left[\exp\left(\delta(ph) - \frac{1}{2}\|ph\|_{\mathcal{H}}^2 \right) \right] \exp\left(\frac{p^2 - p}{2}\|h\|_{\mathcal{H}}^2 \right)
$$

$$
= \exp\left(\frac{p^2 - p}{2}\|h\|_{\mathcal{H}}^2 \right).
$$

Hence,

$$
\mathbf{E}\left[\left| \exp'_M \left(\delta h - \frac{1}{2}\|h\|_{\mathcal{H}}^2 \right) \right|^p \right] \leq \exp\left(\frac{p^2 - 1}{2}\|h\|_{\mathcal{H}}^2 \right) \frac{M}{\sqrt{2\pi}} \int_{\mathbf{R}} e^{-py} e^{-M^2 y^2/2} dy
$$

$$
= \exp\left(\frac{p^2 - p}{2}\|h\|_{\mathcal{H}}^2 \right) \mathbf{E}\left[\exp\left(-p\mathcal{N}(0, 1/M^2) \right) \right]
$$

$$
= \exp\left(\frac{p^2 - p}{2}\|h\|_{\mathcal{H}}^2 \right) \exp(p/M^2).
$$

Then, (2.11) holds true and the result follows from (2.10) and Lemma 2.3.

Lazy Student Trick

Using the theory of distributions on Wiener space (see [7]), we can almost prove that a functional is differentiable because we know how to compute its

(continued)

derivative. Indeed, F is always differentiable in the sense of distributions and it remains to prove that it defines an element of $L^p(W \to \mathcal{H}; \mu) = L^q(W \to \mathcal{H}; \mu)^*$ to be able to claim that it belongs to $\mathbb{D}_{p,1}$.

The previous proof would then boil down to say that formally

$$\nabla \Lambda_h(\omega) = \Lambda_h(\omega) h$$

and then use the same computations as above to show that

$$\mathbf{E}\left[\|\Lambda_h \, h\|_{\mathcal{H}}^p\right] = \mathbf{E}\left[\Lambda_h^p\right] \|h\|_{\mathcal{H}}^p = \exp\left(\frac{p^2 - p}{2}\|h\|_{\mathcal{H}}^2\right) \|h\|_{\mathcal{H}}^p < \infty,$$

hence $\Lambda_h \in \mathbb{D}_{p,1}$.

Corollary 2.2 *Let F belong to $\mathbb{D}_{p,1}$ and G to $\mathbb{D}_{q,1}$ with $q = p/(p-1)$. If $h \in \mathcal{H}$, then Eq. (2.5) holds*

$$\mathbf{E}\left[G \langle \nabla F, h \rangle_{\mathcal{H}}\right] = -\mathbf{E}\left[F \langle \nabla G, h \rangle_{\mathcal{H}}\right] + \mathbf{E}\left[F G \, \delta h\right].$$

Proof According to Lemma 2.2, it is true for F and G in $\mathcal{S}$. Let $(F_n, n \geq 0)$ be a sequence of elements of $\mathcal{S}$ converging to F in $\mathbb{D}_{p,1}$. Since G belongs to $\mathcal{S}$, G and $\nabla_h G$ belong to $L^q(W \to \mathbf{R}; \mu)$. By Hölder inequality, we see that (2.5) holds for $F \in \mathbb{D}_{p,1}$ and $G \in \mathcal{S}$. Repeat the same approach with an approximation of $G \in \mathbb{D}_{q,1}$ by elements of $\mathcal{S}$. $\qquad\Box$

We can now generalize the basic formulas to elements of $\mathbb{D}_{p,1}$ whose proof are obtained by density.

Theorem 2.4 *For $F \in \mathbb{D}_{p,1}$ and $G \in \mathbb{D}_{q,1}$ (with $1/p + 1/q = 1/r$ for $r > 1$), for $\phi \in C_b^1$, the product FG belongs to $\mathbb{D}_{r,1}$ and*

$$\nabla(FG) = F \nabla G + G \nabla F$$

$$\nabla \phi(F) = \phi'(F) \nabla F.$$

More generally, for $U : W \to X$ where X is Banach space, we can reproduce the whole machinery to define its gradient.

Definition 2.5 Consider the X-valued cylindrical functions of the form

$$U(\omega) = f(\delta h_1(\omega), \cdots, \delta h_n) \, x$$

where the first term is an element of S and x is a deterministic element of X. Then, define ∇F as the element of $L^p(W \to \mathcal{H} \otimes X;\ \mu)$ given by

$$\nabla U(\omega) = \sum_{j=1}^{n} \partial_j f(\delta h_1(\omega), \cdots, \delta h_n)\, h_j \otimes x$$

and consider $\mathbb{D}_{p,1}(X)$ the completion of the vector space of X-valued cylindrical functions with respect to the norm

$$\|U\|_{\mathbb{D}_{p,1}(X)} = \|U\|_{L^p(W \to X;\ \mu)} + \|\nabla U\|_{L^p(W \to \mathcal{H} \otimes X;\ \mu)}.$$

> **Support of the Gradient and Adaptability**
> The Malliavin calculus does not need any notion of time to be developed. The definition of the gradient relies only on the properties of the Gaussian measure, which can be defined for processes indexed by several variables like the Brownian sheet. It is then remarkable that, in the end, there exists a link between measurability and support of the gradient.

We need to introduce the two families of projections:

Definition 2.6 For any $t \in [0, 1]$, we set

$$\dot{\pi}_t\ :\ L^2([0, 1] \to \mathbf{R};\ \ell) \longrightarrow L^2([0, 1] \to \mathbf{R};\ \ell)$$

$$\dot{h} \longmapsto \dot{h}\mathbf{1}_{[0,t]},$$

$$\pi_t\ :\ \mathcal{H} \longrightarrow \mathcal{H}$$

$$h = I^1(\dot{h}) \longmapsto I^1(\dot{h}\mathbf{1}_{[0,t]}).$$

We have

$$\|\pi_t h\|_{\mathcal{H}}^2 = \int_0^1 \dot{h}(s)^2 \mathbf{1}_{[0,t]}(s)\mathrm{d}s \le \|\dot{h}\|_{L^2}^2 = \|h\|_{\mathcal{H}}^2,$$

meaning that π_t is continuous on $\mathcal{H}$. Moreover,

$$\pi_t(s \wedge .) = I^1\big(\dot{\pi}_t(\mathbf{1}_{[0,s]})\big) = I^1\big(\mathbf{1}_{[0,s]}\mathbf{1}_{[0,t]}\big) = I^1\big(\mathbf{1}_{[0,s \wedge t]}\big) = (t \wedge s)\wedge$$

so that

$$\pi_t(s \wedge .) = \begin{cases} s\wedge, & \text{if } s \le t \\ t\wedge, & \text{otherwise.} \end{cases} \tag{2.12}$$

Lemma 2.4 *Let $F \in \mathbb{D}_{p,1}$ and $\mathcal{F}_t = \sigma\{\omega(s), \ s \leq t\}$. Then, $\mathbf{E}[F \,|\, \mathcal{F}_t]$ belongs to $\mathbb{D}_{p,1}$ and we have*

$$\pi_t \mathbf{E}[\nabla F \,|\, \mathcal{F}_t] = \nabla \mathbf{E}[F \,|\, \mathcal{F}_t] \tag{2.13}$$

Furthermore, if F is $\mathcal{F}_t$-measurable, then $\dot{\nabla}_s F = 0$ for all $s > t$.

Proof **Step 1** First consider that F is cylindrical. For the sake of simplicity, imagine that

$$F = f\Big(B(t_1), B(t_2)\Big) \text{ with } t_1 < t < t_2.$$

Then,

$$
\begin{aligned}
\mathbf{E}[F \,|\, \mathcal{F}_t] &= \mathbf{E}\left[f\Big(B(t_1), B(t_2) - B(t) + B(t)\Big)\right] \\
&= \int_{\mathbf{R}} f\Big(B(t_1), B(t) + x\Big) p_{t_2-t}(x)\mathrm{d}x \\
&= \tilde{f}\Big(B(t_1), B(t)\Big),
\end{aligned}
\tag{2.14}
$$

where p_{t_2-t} is the density of $B(t_2) - B(t)$, i.e., of a centered Gaussian distribution of variance $(t_2 - t)$ and

$$\tilde{f}(u, v) = \int_{\mathbf{R}} f\big(u, v + x\big) p_{t_2-t}(x)\mathrm{d}x \text{ belongs to Schwartz}(\mathbf{R}^2).$$

On the one hand,

$$\nabla_s \mathbf{E}[F \,|\, \mathcal{F}_t] = \partial_1 \tilde{f}\big(B(t_1), B(t)\big) t_1 \wedge s + \partial_2 \tilde{f}\big(B(t_1), B(t)\big) t \wedge s. \tag{2.15}$$

On the other hand,

$$
\begin{aligned}
\mathbf{E}[\nabla_s F \,|\, \mathcal{F}_t] = \mathbf{E}\big[\partial_1 f\big(B(t_1), B(t_2)\big) \,|\, \mathcal{F}_t\big] \, t_1 \wedge s \\
+ \mathbf{E}\big[\partial_2 f\big(B(t_1), B(t_2)\big) \,|\, \mathcal{F}_t\big] \, t_2 \wedge s.
\end{aligned}
\tag{2.16}
$$

The same reasoning as in (2.14) leads to

$$
\begin{aligned}
\mathbf{E}[\partial_i f(B(t_1), B(t_2)) \,|\, \mathcal{F}_t] &= \int_{\mathbf{R}} \partial_i f\big(B(t_1), B(t) + x\big) p_{t-t_2}(x)\mathrm{d}x \\
&= \partial_i \tilde{f}\big(B(t_1), B(t)\big),
\end{aligned}
\tag{2.17}
$$

for $i \in \{1, 2\}$. In view of (2.17), Eq. (2.16) becomes

$$\mathbf{E}[\nabla_s F \mid \mathcal{F}_t] = \sum_{i=1}^{2} \partial_i \tilde{f}(B(t_1), B(t)) \, t_i \wedge s. \tag{2.18}$$

Thus, according to (2.12),

$$\pi_t \mathbf{E}[\nabla_s F \mid \mathcal{F}_t] = \sum_{i=1}^{2} \partial_i \tilde{f}(B(t_1), B(t)) \, \pi_t(t_i \wedge .)(s)$$

$$= \sum_{i=1}^{2} \partial_i \tilde{f}(B(t_1), B(t)) \, (t_i \wedge t) \wedge s$$

$$= \nabla_s \mathbf{E}[F \mid \mathcal{F}_t].$$

Step 2 For the general case, let $(F_n, n \geq 0)$ be a sequence of elements of $\mathcal{S}$ converging to F in $\mathbb{D}_{p,1}$. We can construct a sequence of cylindrical functions that are $\mathcal{F}_t$ measurable and converge in $\mathbb{D}_{p,1}$ to $\mathbf{E}[F \mid \mathcal{F}_t]$. For any n, there exist $t_1^n < \ldots < t_{k_n}^n$ such that $F_n = f_n(B(t_1^n), \cdots, B(t_{k_n}^n))$. If $t_{j_0}^n \leq t < t_{j_0+1}^n$, for $l \geq j_0 + 1$, replace $B(t_l^n)$ by

$$\left(B(t_l^n) - B(t_{l-1}^n)\right) + \ldots + \left(B(t_{j_0+1}^n) - B(t)\right) + B(t).$$

Let W^n be the Gaussian vector whose coordinates are the independent Gaussian random variables $(B(t_{k_n}^n) - B(t_{k_n-1}^n)), \cdots, B(t_{j_0+1}^n) - B(t))$ and

$$\kappa_n : \mathbf{R}^{k_n} \longrightarrow \mathbf{R}^{k_n}$$

$$w = (w_i, \, 1 \leq i \leq k_n) \longmapsto w_i \text{ if } i \leq j_0,$$

$$\longmapsto w_i + B(t) + \sum_{l=1}^{i-j_0} W_l^n \text{ if } i > j_0.$$

Hence

$$\mathbf{E}[F_n \mid \mathcal{F}_t] = \mathbf{E}\left[(f_n \circ \kappa_n)(B(t_1^n), \cdots, B(t_{j_0}^n)) \mid B(t_1^n), \cdots, B(t)\right].$$

Starting from this identity, we can reproduce the latter reasoning and see that (2.13) holds for such functionals.

Step 3 It remains to prove that $\mathbf{E}[F_n \mid \mathcal{F}_t]$ converges to $F = \mathbf{E}[F \mid \mathcal{F}_t]$ in $\mathbb{D}_{p,1}$. By Jensen inequality,

$$\mathbf{E}\left[|\mathbf{E}[F_n \mid \mathcal{F}_t] - \mathbf{E}[F \mid \mathcal{F}_t]|^p\right] \leq \mathbf{E}\left[|F_n - F|^p\right] \xrightarrow{n \to \infty} 0.$$

According to Proposition 2.1, the dual of $L^p(W \to \mathcal{H}; \mu)$ is $L^q(W \to \mathcal{H}; \mu)$ and

$$\|\nabla \mathbf{E}[F_n \mid \mathcal{F}_t] - \nabla \mathbf{E}[F_m \mid \mathcal{F}_t]\|_{L^p(W \to \mathcal{H}; \mu)}$$

$$= \sup\left\{\left|\mathbf{E}\left[\langle \nabla \mathbf{E}[F_n \mid \mathcal{F}_t] - \nabla \mathbf{E}[F_m \mid \mathcal{F}_t], h\rangle_{\mathcal{H}} \ G\right]\right|, \ \|h\|_{\mathcal{H}} = 1, \ \|G\|_{L^q} = 1\right\}.$$

Then, (2.13) implies that

$$\left|\mathbf{E}\left[\langle \nabla \mathbf{E}[F_n \mid \mathcal{F}_t] - \nabla \mathbf{E}[F_m \mid \mathcal{F}_t], h\rangle_{\mathcal{H}} \ G\right]\right|$$

$$= \left|\mathbf{E}\left[\langle \pi_t \mathbf{E}[\nabla(F_n - F_m) \mid \mathcal{F}_t], h\rangle_{\mathcal{H}} \ G\right]\right|$$

$$= \left|\mathbf{E}\left[\langle \mathbf{E}[\nabla(F_n - F_m) \mid \mathcal{F}_t], \pi_t h\rangle_{\mathcal{H}} \ G\right]\right|$$

$$\leq \|\nabla(F_n - F_m)\|_{L^p(W \to \mathcal{H}; \mu)} \|h\|_{\mathcal{H}} \|G\|_{L^q(W \to \mathbf{R}; \mu)}.$$

Since $(\nabla F_n, \ n \geq 0)$ is a Cauchy sequence in $L^p(W \to \mathcal{H}; \mu)$, so does the sequence $(\nabla \mathbf{E}[F_n \mid \mathcal{F}_t], \ n \geq 0)$, hence it is a converging sequence. Since ∇ is closable, the limit can only be $\nabla \mathbf{E}[F \mid \mathcal{F}_t]$.

Step 4 Recall that ε_s is the Dirac mass at point s and that $\varepsilon_s \in W^*$. Let $H_t^\perp = \bigcap_{s \in [t,1] \cap \mathbf{Q}} \ker(\varepsilon_s - \varepsilon_t)$; it is a denumerable intersection of closed subspaces of $\mathcal{H}$, hence it is closed in $\mathcal{H}$. By sample-paths continuity of the elements of $\mathcal{H}$, $\dot{h}(s) = 0$ for $s > t$ means that $h(s) = h(t)$ for any $s > t$ and $s \in \mathbf{Q}$, which is equivalent to $h \in H_t^\perp$. From Step 3, we know that there exists a subsequence; we still denote by $(F_n, n \geq 0)$, such that $\nabla \mathbf{E}[F_n \mid \mathcal{F}_t]$ converges almost-surely in $\mathcal{H}$ to $\nabla \mathbf{E}[F \mid \mathcal{F}_t]$. From Step 2, we know that for any $n \geq 1$, $\nabla \mathbf{E}[F_n \mid \mathcal{F}_t]$ belongs to $H_t^\perp$. Since $H_t^\perp$ is closed, $\nabla \mathbf{E}[F \mid \mathcal{F}_t]$ belongs to $H_t^\perp$.

$\square$

As we saw above, an element U of $L^p(W \to \mathcal{H}; \mu)$ can be represented as

$$U(\omega, t) = \int_0^t \dot{U}(\omega, s)\mathrm{d}s, \ \text{for all } t \in [0, 1], \tag{2.19}$$

where $\dot{U}$ is measurable from $W \times [0, 1]$ onto $\mathbf{R}$.

Definition 2.7 An $\mathcal{H}$-valued random variable U is said to be adapted whenever the process $\dot{U}$ given by (2.19) is adapted in the classical sense.

We denote by $L_a^2(W \to \mathcal{H}; \mu)$ the set of $\mathcal{H}$-valued adapted, random variables such that

$$\mathbf{E}\left[\int_0^1 |\dot{U}(s)|^2 \mathrm{d}s\right] = \mathbf{E}\left[\|U\|_{\mathcal{H}}^2\right] < \infty.$$

It is a closed subspace of $L^2(W \to \mathcal{H}; \mu)$: For a sequence of adapted processes that converges to some process in $L^2(W \to \mathcal{H}; \mu)$, there exists a subsequence that converges with probability 1 hence the adaptability is transferred to the limiting process.

Similarly $\mathbb{D}_{2,1}^a(\mathcal{H})$ is the subset of $L_a^2(W \to \mathcal{H}; \mu)$ such that

$$\mathbf{E}\left[\iint |\dot{\nabla}_r \dot{U}(s)|^2 dr ds\right] = \|\nabla U\|_{L^2(W \to \mathcal{H}; \mu)}^2 < \infty.$$

Theorem 2.5 *Let U belong to $\mathbb{D}_{2,1}^a(\mathcal{H})$ and $\mathcal{D}_n$ be the dyadic partition of $[0, 1]$ of step 2^{-n}. Then,*

$$\dot{U}_{\mathcal{D}_n}(t) = \sum_{i=1}^{2^n-1} 2^n \left(\int_{(i-1)2^{-n}}^{i\,2^{-n}} \dot{U}(r) dr\right) \mathbf{1}_{(i2^{-n},(i+1)2^{-n}]}(t) \qquad (2.20)$$

converges in $\mathbb{D}_{2,1}^a(\mathcal{H})$ to U.

Proof **Step 1** Since indicator functions with disjoint support are orthogonal in $L^2([0, 1] \to \mathbf{R}; \ell)$, we have

$$\int_0^1 |\dot{U}_{\mathcal{D}_n}(t)|^2 dt = \sum_{i=1}^{2^n-1} \left(2^n \int_{(i-1)2^{-n}}^{i\,2^{-n}} \dot{U}(r) dr\right)^2 \int_0^1 \mathbf{1}_{(i2^{-n},(i+1)2^{-n}]}(t) dt$$

$$\leq \sum_{i=1}^{2^n-1} \int_{(i-1)2^{-n}}^{i\,2^{-n}} |\dot{U}(r)|^2 \frac{dr}{2^{-n}} 2^{-n} = \int_0^1 |\dot{U}(r)|^2 dr,$$

according to the Jensen inequality. Hence,

$$\mathbf{E}\left[\int_0^1 |\dot{U}_{\mathcal{D}_n}(t)|^2 dt\right] \leq \mathbf{E}\left[\int_0^1 |\dot{U}(r)|^2 dr\right].$$

In other words, this means that the maps

$$p_{\mathcal{D}_n} : L^2(W \to \mathcal{H}; \mu) \longrightarrow L^2(W \to \mathcal{H}; \mu)$$

$$U \longmapsto I^1(\dot{U}_{\mathcal{D}_n})$$

are continuous and satisfy

$$\|p_{\mathcal{D}_n}\| \leq 1. \qquad (2.21)$$

Let

$$M = \left\{U \in L^2(W \to \mathcal{H}; \mu), \dot{U} \text{ is a.s. continuous and } \mathbf{E}\left[\|\dot{U}\|_\infty^2\right] < \infty\right\}.$$

For such a process

$$\|\dot{U} - \dot{U}_{\mathcal{D}_n}\|^2_{L^2([0,1] \to \mathbf{R};\, \ell)} \le \sum_{i=1}^{2^n-1} \int_{(i-1)2^{-n}}^{i\,2^{-n}} \left(2^n \int_{(i-1)2^{-n}}^{i\,2^{-n}} |\dot{U}(r) - \dot{U}(t)| dr \right)^2 dt$$

$$\le \sum_{i=1}^{2^n-1} \int_{(i-1)2^{-n}}^{i\,2^{-n}} 2^n \int_{(i-1)2^{-n}}^{i\,2^{-n}} |\dot{U}(r) - \dot{U}(t)|^2 dr\, dt,$$

by the Jensen inequality. Since $\dot{U}$ is a.s. continuous, for $t \in ((i-1)2^{-n}, \, i\,2^{-n}]$,

$$2^n \int_{(i-1)2^{-n}}^{i\,2^{-n}} |\dot{U}(r) - \dot{U}(t)|^2 dr \xrightarrow[\text{a.s.}]{n \to \infty} 0.$$

Since $\mathbf{E}\left[\|\dot{U}\|^2_\infty\right]$ is finite, the dominated convergence theorem entails that

$$\mathbf{E}\left[\sum_{i=1}^{2^n-1} \int_{(i-1)2^{-n}}^{i\,2^{-n}} 2^n \int_{(i-1)2^{-n}}^{i\,2^{-n}} |\dot{U}(r) - \dot{U}(t)|^2 dr\, dt \right] \xrightarrow{n \to \infty} 0.$$

We have thus proved that for $U \in M$, $p_{\mathcal{D}_n} U$ converges to U in $L^2(\mathbf{W} \otimes [0, 1] \to \mathbf{R};\, \mu \otimes \ell)$. For $U \in L^2(\mathbf{W} \to \mathcal{H};\, \mu)$, for any $\epsilon > 0$, there exists $U_\epsilon \in M$ such that

$$\|U - U_\epsilon\|_{L^2(\mathbf{W} \to \mathcal{H};\, \mu)} \le \epsilon.$$

In view of (2.21),

$$\|U - p_{\mathcal{D}_n} U\|_{L^2(\mathbf{W} \to \mathcal{H};\, \mu)} \le \|U - U_\epsilon\|_{L^2(\mathbf{W} \to \mathcal{H};\, \mu)} + \|p_{\mathcal{D}_n}(U - U_\epsilon)\|_{L^2(\mathbf{W} \to \mathcal{H};\, \mu)}$$

$$+ \|p_{\mathcal{D}_n}(U_\epsilon) - U_\epsilon\|_{L^2(\mathbf{W} \to \mathcal{H};\, \mu)}$$

$$\le 2\|U - U_\epsilon\|_{L^2(\mathbf{W} \to \mathcal{H};\, \mu)} + \|p_{\mathcal{D}_n}(U_\epsilon) - U_\epsilon\|_{L^2(\mathbf{W} \to \mathcal{H};\, \mu)}$$

$$\le 2\epsilon + \|p_{\mathcal{D}_n}(U_\epsilon) - U_\epsilon\|_{L^2(\mathbf{W} \to \mathcal{H};\, \mu)}.$$

It remains to choose n sufficiently large to have the rightmost term less than ϵ to prove that $\dot{U}_{\mathcal{D}_n}$ tends to $\dot{U}$ in $L^2(\mathbf{W} \otimes [0, 1] \to \mathbf{R};\, \mu \otimes \ell)$.

Step 2 Remark that if $\dot{U}$ is adapted, then so does $\dot{U}_{\mathcal{D}_n}$ since we chose carefully the interval of the integral in (2.20).

Step 3 Similarly, if $U \in \mathbb{D}_{2,1}$, $\dot{\nabla}_r \dot{U}_t$ can be approximated in $L^2(W \otimes [0, 1]^2 \to \mathbf{R}; \mu \otimes \ell^{\otimes 2})$ by

$$\sum_{i=1}^{2^n-1} 2^n \left(\int_{(i-1)2^{-n}}^{i \, 2^{-n}} \dot{\nabla}_r \dot{U}(s) \mathrm{d}s \right) \mathbf{1}_{(i2^{-n},(i+1)2^{-n}]}(t).$$

Then, the same proof as before shows this approximation converges in the space $L^2(W \otimes [0, 1]^2 \to \mathbf{R}; \mu \otimes \ell^{\otimes 2})$ to $\dot{\nabla}\dot{U}$.

$\square$

This approximation is necessary to compute the derivative of an Itô integral. It is the analog of the usual formula

$$\frac{d}{d\tau} \left(\int_0^\tau f(\tau, s) \mathrm{d}s \right) = f(\tau, \tau) + \int_0^\tau \frac{\partial f}{\partial \tau}(\tau, s) \mathrm{d}s,$$

since we have some ω's both in $\dot{U}$ and in dB.

Theorem 2.6 *For $U \in \mathbb{D}_{2,1}^a(\mathcal{H})$, the Itô integral of $\dot{U}$ belongs to $\mathbb{D}_{2,1}$ and for any $h \in \mathcal{H}$,*

$$\left\langle \nabla\left(\int \dot{U}(s) \mathrm{d}B(s) \right), h \right\rangle_{\mathcal{H}} = \int_0^1 \dot{U}(s) \dot{h}(s) \mathrm{d}s + \int_0^1 \langle \nabla \dot{U}(s), h \rangle_{\mathcal{H}} \mathrm{d}B(s). \tag{2.22}$$

Proof From the previous theorem, we know that $\langle \nabla \dot{U}(s), h \rangle_{\mathcal{H}}$ is adapted and square integrable so that its stochastic integral is well defined. For $U(t) = U_a I^1(\mathbf{1}_{(a,b]})(t)$ with $U_a \in \mathcal{F}_a$ and $U_a \in \mathbb{D}_{2,1}$, on the one hand, since ∇ is a derivation operator, we have

$$\left\langle \nabla\left(\int \dot{U}(s) \mathrm{d}B(s) \right), h \right\rangle_{\mathcal{H}}$$
$$= \left\langle \nabla\left(U_a (B(b) - B(a)) \right), h \right\rangle_{\mathcal{H}}$$
$$= \langle \nabla U_a, h \rangle_{\mathcal{H}} (B(b) - B(a)) + \int_0^1 U_a \mathbf{1}_{(a,b]}(s) \dot{h}(s) \mathrm{d}s$$
$$= \int_0^1 \langle \nabla U_a, h \rangle_{\mathcal{H}} \mathbf{1}_{(a,b]}(s) \mathrm{d}B(s) + \int_0^1 U_a \mathbf{1}_{(a,b]}(s) \dot{h}(s) \mathrm{d}s$$
$$= \int_0^1 \dot{U}(s) \dot{h}(s) \mathrm{d}s + \int_0^1 \langle \nabla \dot{U}(s), h \rangle_{\mathcal{H}} \mathrm{d}B(s).$$

By linearity, Eq. (2.22) holds for simple processes as in Theorem 2.5. Since for U with continuous sample-paths, $U_{\mathcal{D}_n}$ tends in $L^2(W \times [0, 1], \mu \otimes \ell)$ to U, in virtue of Lemma 2.3, it remains to prove that

$$\sup_n \mathbf{E}\left[\| \nabla \int \dot{U}_{\mathcal{D}_n}(s) \mathrm{d}B(s) \|_{\mathcal{H}}^2 \right] < \infty.$$

By the very definition of the Pettis integral,

$$\left\langle \int_0^1 \nabla \dot{U}_{\mathcal{D}_n}(s) \mathrm{d}B(s), \, h \right\rangle_{\mathcal{H}} = \int_0^1 \langle \nabla \dot{U}_{\mathcal{D}_n}(s), \, h \rangle_{\mathcal{H}} \mathrm{d}B(s).$$

In view of (2.22), the hard part is then to show that

$$\sup_n \mathbf{E} \left[\| \int_0^1 \nabla \dot{U}_{\mathcal{D}_n}(s) \mathrm{d}B(s) \|_{\mathcal{H}}^2 \right] < \infty.$$

We remark that

$$t \longmapsto \int_0^t \nabla \dot{U}_{\mathcal{D}_n}(s) \mathrm{d}B(s)$$

is an Hilbert valued martingale and we admit that the Itô isometry is still valid:

$$\mathbf{E} \left[\left| \int_0^1 \nabla \dot{U}_{\mathcal{D}_n}(s) \mathrm{d}B(s) \right|^2 \right] = \mathbf{E} \left[\int_0^1 \| \nabla \dot{U}_{\mathcal{D}_n}(s) \|_{\mathcal{H}}^2 \mathrm{d}s \right]$$

$$= \mathbf{E} \left[\int_0^1 \int_0^1 |\dot{\nabla}_r \dot{U}_{\mathcal{D}_n}(s)|^2 \mathrm{d}r \mathrm{d}s \right] = \| \nabla U_{\mathcal{D}_n} \|_{L^2(\mathbf{W} \to \mathcal{H} \otimes \mathcal{H}; \, \mu)}^2.$$

Combining (2.22) with this upper-bound, we get

$$\mathbf{E} \left[\| \nabla \int \dot{U}_{\mathcal{D}_n}(s) \mathrm{d}B(s) \|_{\mathcal{H}}^2 \right] \leq 2 \left(\| U_{\mathcal{D}_n} \|_{L^2(\mathbf{W} \to \mathcal{H}; \, \mu)}^2 + \| \nabla U_{\mathcal{D}_n} \|_{L^2(\mathbf{W} \to \mathcal{H} \otimes \mathcal{H}; \, \mu)}^2 \right).$$

We conclude with Theorem 2.5.   □

For cylindrical functions, we can clearly define higher order derivative following the same rule. The only difficulty is to realize that the second (respectively k-th) order gradient belongs to $\mathcal{H}^{\otimes(2)}$ (respectively $\mathcal{H}^{\otimes(k)}$): For instance, for $F = f(\delta h_j, \, 1 \leq j \leq n)$,

$$\left\langle \nabla^{(2)} F, \, h \otimes k \right\rangle_{\mathcal{H} \otimes \mathcal{H}} = \sum_{j,l=1}^n \partial_{j,l} f(\delta h_j, \, 1 \leq j \leq n) \, \langle h_j, \, h \rangle_{\mathcal{H}} \otimes \langle h_l, \, k \rangle_{\mathcal{H}}$$

$$= \left\langle \nabla(\langle \nabla F, h \rangle_{\mathcal{H}}), \, k \right\rangle_{\mathcal{H}}.$$

Definition 2.8 For any $p > 1$ and $k \geq 1$, $\mathbb{D}_{p,k}$ is the completion of $\mathcal{S}$ with respect to the norm

$$\| F \|_{p,k} = \| F \|_p + \sum_{j=1}^k \| \nabla^{(j)} F \|_{L^p(\mathbf{W} \to \mathcal{H}^{\otimes(j)}; \, \mu)}.$$

The space of *test functions* is $\mathbb{D} = \cap_{p>1} \cap_{k \geq 1} \mathbb{D}_{p,k}$. It plays the same rôle as the set of C^∞ functions with compact support plays in the theory of distributions.

Example (Second Derivative of $B(t)^2$) We know that

$$\nabla B(t)^2 = 2 B(t) \nabla B(t) = 2 B(t) t \wedge .$$

By iteration,

$$\nabla^{(2)} B(t)^2 = 2 \nabla B(t) \otimes t \wedge . = 2 (t \wedge .) \otimes (t \wedge .)$$

or equivalently

$$\nabla^{(2)}_{r,s} B(t)^2 = 2 (t \wedge r) (t \wedge s).$$

Is $\nabla^{(2)} F$ a Map from $\mathcal{H}$ to $\mathcal{H}$ or a Function of Two Parameters ?
By its very definition $\nabla^{(2)} F(\omega)$ is an element of $\mathcal{H} \otimes \mathcal{H}$ that is to say a continuous linear form on $\mathcal{H} \times \mathcal{H}$, i.e., it takes as its argument $h, k \in \mathcal{H}$ and yields a real number. Alternatively, we can consider the map

$$\mathcal{H} \longrightarrow \mathcal{H}^* \simeq \mathcal{H}$$

$$h \longmapsto \left(k \mapsto \langle \nabla^{(2)} F(\omega), h \otimes k \rangle_{\mathcal{H} \otimes \mathcal{H}} \right).$$

As such $\nabla^{(2)} F(\omega)$ appears as a linear map from $\mathcal{H}$ into itself. To make things even more confusing, we can work with the L^2 representatives. By the very construction of tensor products, it is immediate that

$$I^1 \otimes I^1 : L^2([0, 1]^2 \to \mathbf{R}; \ell) \simeq L^2([0, 1] \to \mathbf{R}; \ell)^{\otimes(2)} \longrightarrow \mathcal{H} \otimes \mathcal{H}$$

$$\dot{h} \otimes \dot{k} \longmapsto I^1(\dot{h}) \otimes I^1(\dot{k})$$

can be extended in a bijective isometry and we denote by $\dot{\nabla}^{(2)} F(\omega)$ the pre-image of $\nabla^{(2)} F(\omega)$ by this map so that we have

$$\langle \nabla^{(2)} F(\omega), h \otimes k \rangle_{\mathcal{H} \otimes \mathcal{H}} = \int_0^1 \int_0^1 \dot{\nabla}^{(2)}_{s,r} F(\omega) \dot{h}(s) \dot{k}(r) ds dr. \qquad (2.23)$$

In ordinary differential calculus, the Schwarz Theorem says that the order of differentiation is unimportant. The analog here is the say that $\nabla^{(2)}$ is a symmetric operator.

Lemma 2.5 *Assume that some* $p \geq 1$, $F \in \mathbb{D}_{p,2}$. *Then, for any* $k, l \in \mathcal{H}$,

$$\left\langle \nabla^{(2)} F, \, k \otimes l \right\rangle_{\mathcal{H} \otimes \mathcal{H}} = \left\langle \nabla^{(2)} F, \, l \otimes k \right\rangle_{\mathcal{H} \otimes \mathcal{H}}$$

Moreover, for $F \in \mathbb{D}_{2,2}$, *for almost all* $\omega \in W$, $\nabla^{(2)} F(\omega)$ *viewed as a map from* $\mathcal{H}$ *into itself is Hilbert–Schmidt.*

***Proof* Step 1** $F \in \mathbb{D}_{2,2}$ means that

$$\infty > \mathbf{E}\left[\|\nabla^{(2)} F\|_{\mathcal{H}}^2 \right] = \mathbf{E}\left[\int_0^1 \int_0^1 |\dot{\nabla}_r \dot{\nabla}_s F(\omega)|^2 dr ds \right].$$

This implies that almost all $\omega \in W$,

$$\int_0^1 \int_0^1 |\dot{\nabla}_r \dot{\nabla}_s F(\omega)|^2 dr ds < \infty,$$

which in view of (2.23) means that $\nabla^{(2)} F(\omega)$ is Hilbert–Schmidt.

Step 2 For $F \in \mathcal{S}$, $F = f(\delta h_1, \cdots, \delta h_M)$, in virtue of the Schwarz theorem for crossed derivatives of functions of several variables,

$$\left\langle \nabla^{(2)} F, \, k \otimes l \right\rangle_{\mathcal{H} \otimes \mathcal{H}} = \sum_{i,j=1}^n \partial_{ij}^2 f(\delta h_1, \cdots, \delta h_M) \, \langle k, h_i \rangle_{\mathcal{H}} \langle l, h_j \rangle_{\mathcal{H}}$$

$$= \sum_{i,j=1}^n \partial_{ji}^2 f(\delta h_1, \cdots, \delta h_M) \, \langle l, h_j \rangle_{\mathcal{H}} \langle k, h_i \rangle_{\mathcal{H}}$$

$$= \left\langle \nabla^{(2)} F, \, l \otimes k \right\rangle_{\mathcal{H} \otimes \mathcal{H}}.$$

Furthermore, (2.23) entails that

$$\left| \mathbf{E}\left[\left\langle \nabla^{(2)} F, \, k \otimes l - l \otimes k \right\rangle_{\mathcal{H} \otimes \mathcal{H}} \right] \right| \leq 2 \|D\|_{\mathbb{D}_{2,2}} \|k\|_{\mathcal{H}} \|l\|_{\mathcal{H}},$$

hence the proof by density of $\mathcal{S}$ in $\mathbb{D}_{2,2}$.

$\square$

2.2 Divergence

For a matrix $M \in \mathcal{M}_{n,p}(\mathbf{R})$, its adjoint, which turns to coincide with its transpose, is defined by the identity:

$$\langle Mx, \, y \rangle_{\mathbf{R}^p} = \langle x, \, M^* y \rangle_{\mathbf{R}^n}.$$

We see that to define an adjoint, we need to have a notion a scalar product or more generally of a duality bracket. It is then natural for M continuous from a Banach E into a Banach F to define its adjoint as the continuous map from F^* into E^* defined by the identity:

$$\langle Mx, y \rangle_{F,F^*} = \langle x, M^*y \rangle_{E,E^*}.$$

For any $q > 1$, the Gross–Sobolev derivative, which we denoted by ∇, is continuous between the two spaces:

$$\mathbb{D}_{q,1} \subset L^q(W \to \mathbf{R}; \mu) \longrightarrow L^q(W \to \mathcal{H}; \mu).$$

Therefore its adjoint is a map from

$$\left(L^q(W \to \mathcal{H}; \mu)\right)^* = L^p(W \to \mathcal{H}; \mu)$$

$$\longrightarrow \left(L^q(W \to \mathbf{R}; \mu)\right)^* = L^p(W \to \mathbf{R}; \mu)$$

with $1/p + 1/q = 1$ and must satisfy the identity

$$\langle \nabla F, U \rangle_{L^q(W \to \mathcal{H}; \mu), L^p(W \to \mathcal{H}; \mu)} = \langle F, \nabla^* U \rangle_{L^q(W \to \mathbf{R}; \mu), L^p(W \to \mathbf{R}; \mu)}$$

$$\iff \mathbf{E}[\langle \nabla F, U \rangle_{\mathcal{H}}] = \mathbf{E}\left[F \nabla^* U\right].$$

An additional difficulty comes from the fact that ∇ is not defined on the whole of $L^q(W \to \mathbf{R}; \mu)$ but only on the subset $\mathbb{D}_{q,1}$, hence we need to take some restrictions in the definition of the adjoint.

Definition 2.9 Let $p > 1$. Let $\mathrm{Dom}_p \nabla^*$ be the set of $\mathcal{H}$-valued random variables U for which there exists $c_p(U)$ such that for any $F \in \mathbb{D}_{q,1}$,

$$\left|\mathbf{E}\left[\langle \nabla F, U \rangle_{\mathcal{H}}\right]\right| \le c_p(U) \|F\|_{L^q(W \to \mathbf{R}; \mu)}.$$

In this case, we define $\nabla^* U$ as the unique element of $L^p(W \to \mathbf{R}; \mu)$ such that

$$\mathbf{E}\left[\langle \nabla F, U \rangle_{\mathcal{H}}\right] = \mathbf{E}\left[F \nabla^* U\right].$$

Remark 2.1 (∇^* Coincides with the Wiener Integral on $\mathcal{H}$) Recall that δ is the Wiener integral. We now show that $\delta = \nabla^*|_{\mathcal{H}}$. For any $F \in S$, according to (2.5), we have

$$\mathbf{E}[\langle \nabla F, h \rangle_{\mathcal{H}}] = \mathbf{E}[F \delta h] \qquad (2.24)$$

and δh is a Gaussian random variable of variance $\|h\|_{\mathcal{H}}^2$, thus belongs to any $L^q(\mathbf{W} \to \mathbf{R}; \mu)$ for any $q > 1$. Hence,

$$|\mathbf{E}[\langle \nabla F, h \rangle_{\mathcal{H}}]| \leq \|h\|_{\mathcal{H}} \|F\|_{L^p(\mathbf{W} \to \mathbf{R}; \mu)}.$$

This means that h belongs to $\mathrm{Dom}_p \nabla^*$ and (2.24) entails that $\nabla^* h = \delta h$. Henceforth, in the following, we will use the notation δ instead of ∇^* and we keep for further reference the fundamental formula

$$\mathbf{E}[\langle \nabla F, U \rangle_{\mathcal{H}}] = \mathbf{E}[F \, \delta U] \tag{2.25}$$

for any $F \in \mathbb{D}_{q,1}$ and $U \in \mathrm{Dom}_p \, \delta$.

In usual deterministic calculus, if a is a constant, then trivially

$$\int au(s)\mathrm{d}s = a \int u(s)\mathrm{d}s. \tag{2.26}$$

For Itô integrals, this property does not hold any longer since we may have a problem of adaptability: If a is a random variable, not belonging to $\mathcal{F}_0$ and u is an adapted process with all the required integrability properties, then the process $(a\,u(s), \, s \geq 0)$ is not adapted so that $\int au(s)\mathrm{d}B(s)$ is not well defined. For the divergence, since we got rid of the adaptability hypothesis, we can prove a formula analog to (2.26), which is a simple consequence of the fact that ∇ is a derivation operator.

Theorem 2.7 (Divergence of a Process Multiplied by a Random Variable) *Let $U \in \mathrm{Dom}_p \, \delta$ and $a \in \mathbb{D}_{q,1}$ with $1/p + 1/q = 1/r$. Then, $aU \in \mathrm{Dom}_r \, \delta$ and*

$$\delta(aU) = a \, \delta U - \langle \nabla a, U \rangle_{\mathcal{H}}. \tag{2.27}$$

***Proof* Step 1** We first prove that the right-hand side belongs to $L^r(\mathbf{W} \to \mathbf{R}; \mu)$.

$$\mathbf{E}[|a\delta U|^r] \leq \mathbf{E}[|a|^q]^{r/q} \, \mathbf{E}[|\delta U|^p]^{r/p} \tag{2.28}$$

and

$$\begin{aligned} \mathbf{E}[|\langle \nabla a, U \rangle_{\mathcal{H}}|^r] &\leq \mathbf{E}[\|\nabla a\|_{\mathcal{H}}^r \|U\|_{\mathcal{H}}^r] \\ &\leq \mathbf{E}[\|\nabla a\|_{\mathcal{H}}^q]^{r/q} \, \mathbf{E}[\|U\|_{\mathcal{H}}^p]^{r/p} \\ &\leq \|a\|_{\mathbb{D}_{q,1}}^r \|U\|_{\mathbb{D}_{p,1}}^r. \end{aligned} \tag{2.29}$$

Step 2 Denote $r^* = r/(r-1)$. For $F \in \mathbb{D}_{r^*,1}$, since ∇ is a true derivation,

$$\mathbf{E}\left[\langle \nabla F, aU \rangle_{\mathcal{H}}\right] = \mathbf{E}\left[\langle a\nabla F, U \rangle_{\mathcal{H}}\right]$$
$$= \mathbf{E}\left[\langle \nabla(aF) - F\nabla a, U \rangle_{\mathcal{H}}\right] \qquad (2.30)$$
$$= \mathbf{E}\left[F\, a\delta U\right] - \mathbf{E}\left[F\langle \nabla a, U \rangle_{\mathcal{H}}\right].$$

According to (2.28) and (2.29), (2.30) implies that

$$\left|\mathbf{E}\left[\langle \nabla F, aU \rangle_{\mathcal{H}}\right]\right| \le \|a\|_{\mathbb{D}_{p,1}} \|U\|_{\mathbb{D}_{q,1}} \|F\|_{L^{r^*}\left(\mathrm{W} \to \mathcal{H};\, \mu\right)}.$$

Hence, aU belongs to $\mathrm{Dom}_r\, \delta$.

Step 3 At last, (2.30) implies (2.27) by identification.

$\square$

We have already seen that the Itô integral coincides with the Wiener integral for deterministic integrands provided that we identify h and $\dot{h}$. We now show that up to the same identification, the divergence of adapted processes coincides with their Itô integral.

Corollary 2.3 (Divergence Extends Itô Integral) *Let* $U \in \mathbb{D}_{2,1}^a(\mathcal{H})$. *Then,* U *belong to* $\mathrm{Dom}_2\, \delta$ *and*

$$\delta U = \int_0^1 \dot{U}(s)\mathrm{d}B(s), \qquad (2.31)$$

where the stochastic integral is taken in the Itô sense.

Proof The principle of the proof is to establish (2.31) for adapted simple processes and then pass to the limit.

Step 1 For $0 \le s < t \le 1$, let

$$\dot{U}(r) = \theta_s\, \mathbf{1}_{(s,t]}(r), \text{ i.e., } U(r) = \theta_s\left(t \wedge r - s \wedge r\right),$$

where $\theta_s \in \mathbb{D}_{2,1}$ and θ_s is $\mathscr{F}_s$-measurable. According to Theorem 2.7, U is in $\mathrm{Dom}_2\, \delta$ and

$$\delta(U) = \theta_s\, \delta(t \wedge \,.\, - s \wedge \,.) - \langle \nabla\theta_s,\, t \wedge \,.\, - s \wedge \,.\rangle_{\mathcal{H}}$$
$$= \theta_s\left(B(t) - B(s)\right) - \int_0^1 \dot{\nabla}_\tau \theta_s\, \mathbf{1}_{(s,t]}(\tau)\mathrm{d}.\tau \qquad (2.32)$$

Now recall that according to Lemma 2.4, since $\theta_s \in \mathcal{F}_s$,

$$\dot{\nabla}_\tau \theta_s = 0 \text{ if } \tau > r,$$

hence the rightmost integral of (2.32) is null and

$$\delta(U) = \theta_s \left(B(t) - B(s) \right) = \int_0^1 \dot{U}(r) \mathrm{d}B(r). \tag{2.33}$$

Step 2 If $\dot{U}$ is adapted, the random variable

$$2^n \left(\int_{(i-1)2^{-n}}^{i2^{-n}} \dot{U}(r) \mathrm{d}r \right) \text{ belongs to } \mathcal{F}_{i2^{-n}}.$$

Hence, with the notations of Theorem 2.5, we have by linearity

$$\delta(U_{\mathcal{D}_n}) = \int_0^1 \dot{U}_{\mathcal{D}_n}(r) \mathrm{d}B(r).$$

Step 3 It remains to show that we can pass to the limit in both sides of (2.31). The application δ is continuous from $\mathbb{D}_{2,1}^a(\mathcal{H}) \subset \mathbb{D}_{2,1}(\mathcal{H})$ into $L^2(\mathrm{W} \to \mathbf{R};\ \mu)$. Hence, Theorem 2.5 entails that

$$\delta(U_{\mathcal{D}_n}) \xrightarrow[n \to \infty]{L^2(\mathrm{W} \to \mathbf{R};\ \mu)} \delta(U).$$

Furthermore, the Itô integral is an isometry hence a continuous map from $L_a^2(\mathrm{W} \times [0, 1] \to \mathbf{R};\ \mu)$ into $L^2(\mathrm{W} \to \mathbf{R};\ \mu)$. Hence,

$$\int_0^1 \dot{U}_{\mathcal{D}_n}(r) \mathrm{d}B(r) \xrightarrow[n \to \infty]{L^2(\mathrm{W} \to \mathbf{R};\ \mu)} \int_0^1 \dot{U}(r) \mathrm{d}B(r).$$

The proof is thus complete.

$\square$

The Itô isometry states that for U adapted

$$\mathbf{E}\left[\left(\int_0^1 \dot{U}(s) \mathrm{d}B(s) \right)^2 \right] = \mathbf{E}\left[\int_0^1 |\dot{U}(s)|^2 \mathrm{d}s \right].$$

One of the most elegant formula given by the Malliavin calculus is the generalization of this identity to non-adapted integrands.

Remark 2.2 If $U \in \mathbb{D}_{2,1}(\mathcal{H})$, then $\dot{\nabla}\dot{U}$ is a.s. an Hilbert–Schmidt map on $L^2([0,1] \times [0,1], \ell \otimes \ell)$. Indeed, by the definition of the norm in $\mathbb{D}_{2,1}(\mathcal{H})$,

$$
\|U\|_{\mathbb{D}_{2,1}}^2 = \mathbf{E}\left[\|U\|_{\mathcal{H}}^2\right] + \mathbf{E}\left[\|\nabla U\|_{\mathcal{H} \otimes \mathcal{H}}^2\right]
$$
$$
= \mathbf{E}\left[\int_0^1 \dot{U}(s)^2 ds\right] + \mathbf{E}\left[\int_0^1 \int_0^1 |\dot{\nabla}_r \dot{U}(s)|^2 dr ds\right].
$$

This ensures the almost-sure finiteness of

$$
\int_0^1 \int_0^1 |\dot{\nabla}_r \dot{U}(s)|^2 dr ds,
$$

meaning that $\dot{\nabla}\dot{U}$ is Hilbert–Schmidt with probability 1.

Lemma 2.6 *If U belongs to $\mathbb{D}_{2,1}^q(\mathcal{H})$ then* $\mathrm{trace}(\nabla U \circ \nabla U) = 0$.

Proof According to Lemma 1.3,

$$
\mathrm{trace}(\nabla U \circ \nabla U) = \iint_{[0,1]^2} \dot{\nabla}_r \dot{U}(s) \dot{\nabla}_s \dot{U}(r) dr ds.
$$

Since $\dot{U}(s)$ is $\mathcal{F}_s$-measurable, $\dot{\nabla}_r \dot{U}(s) = 0$ if $r > s$. Similarly, $\dot{\nabla}_s \dot{U}(r) = 0$ if $s > r$. Hence, the product is zero $\ell \otimes \ell$ almost-surely. It follows that the integral is null. $\quad\square$

The Itô isometry says that the $L^2(\Omega \to \mathbf{R}; \, \mathbf{P})$-norm of the stochastic integral of an adapted process is equal to the $L^2(\Omega \times [0,1] \to \mathbf{R}; \, \mathbf{P} \otimes \ell)$-norm of the process. The next formula extends this relation to non-adapted integrands and quantify the difference due to non-adaptability.

Theorem 2.8 (L^2 Norm of Divergence) *The space $\mathbb{D}_{1,2}(\mathcal{H})$ is included in $\mathrm{Dom}_2\,\delta$ and for $U \in \mathbb{D}_{1,2}(\mathcal{H})$,*

$$
\mathbf{E}\left[\delta U^2\right] = \mathbf{E}\left[\|U\|_{\mathcal{H}}^2\right] + \mathbf{E}\left[\mathrm{trace}(\nabla U \circ \nabla U)\right]. \tag{2.34}
$$

Lemma 2.7 *For $k \geq 1$, for $V \in \mathbb{D}_{2,1}(\mathcal{H}^{\otimes(k)})$, for $x \in \mathcal{H}^{\otimes(k)}$, for $h \in \mathcal{H}$,*

$$
\langle \nabla \langle V, x \rangle_{\mathcal{H}^{\otimes(k)}}, h \rangle_{\mathcal{H}} = \langle \nabla V, x \otimes h \rangle_{\mathcal{H}^{\otimes(k+1)}}.
$$

Proof For the sake of simplicity, we give the proof for $k = 1$. The general case is handled similarly. Going back to the definition of the scalar product in $\mathcal{H}$, we have

$$
\langle \nabla \langle V, x \rangle_{\mathcal{H}^{\otimes(k)}}, h \rangle_{\mathcal{H}} = \int_0^1 \dot{\nabla}_s \left(\int_0^1 \dot{V}(r) \dot{x}(r) dr \right) \dot{h}(s) ds.
$$

Approximate the inner integral by Riemann sums and pass to the limit to show that

$$\dot{\nabla}_s \left(\int_0^1 \dot{V}(r)\, \dot{x}(r) \mathrm{d}r \right) = \int_0^1 \dot{\nabla}_s \dot{V}(r)\, \dot{x}(r) \mathrm{d}r,$$

first for V such that $(r, s) \longmapsto \dot{\nabla}_s \dot{V}(r)$ is continuous and then by density for all $V \in \mathbb{D}_{2,1}(\mathcal{H})$. Hence the result. □

Proof of Theorem 2.8 For $U \in \mathbb{D}_{1,2}(\mathcal{H})$, U takes its values in $\mathcal{H}$ so that we can write

$$U = \sum_{n \geq 0} \langle U, h_n \rangle_{\mathcal{H}}\, h_n,$$

for $(h_n,\ n \geq 0)$ a complete orthonormal basis of $\mathcal{H}$. The series

$$U_N = \sum_{n=0}^{N} \langle U, h_n \rangle_{\mathcal{H}}\, h_n \ \text{and} \ \nabla U_N = \sum_{n=0}^{N} \nabla \langle U, h_n \rangle_{\mathcal{H}}\, h_n$$

converges in $L^2(\mathrm{W} \to \mathcal{H};\ \mu)$ and $L^2(\mathrm{W} \to \mathcal{H} \otimes \mathcal{H};\ \mu))$, respectively.
According to (2.27),

$$\delta U_N = \sum_{n=0}^{N} \langle U, h_n \rangle_{\mathcal{H}}\, \delta h_n - \sum_{n=0}^{N} \langle \nabla U, h_n \otimes h_n \rangle_{\mathcal{H} \otimes \mathcal{H}}.$$

Thus,

$$\nabla \delta U_N$$

$$= \sum_{n=0}^{N} \left\{ \langle \nabla U, h_n \rangle_{\mathcal{H}}\, \delta h_n + \langle U, h_n \rangle_{\mathcal{H}}\, h_n - \nabla \Big(\langle \nabla U, h_n \otimes h_n \rangle_{\mathcal{H} \otimes \mathcal{H}} \Big) \right\}.$$

$$(2.35)$$

Consequently, in virtue of Lemma 2.7,

$$\mathbf{E}\,[\delta U_N\, \delta U_N] = \sum_{n,k \geq 0}^{N} \mathbf{E}\big[\langle U, h_k \rangle_{\mathcal{H}} \langle \nabla U, h_n \otimes h_k \rangle_{\mathcal{H} \otimes \mathcal{H}}\, \delta h_n \big]$$

$$+ \sum_{n,k \geq 0}^{N} \mathbf{E}\big[\langle U, h_n \rangle_{\mathcal{H}} \langle U, h_k \rangle_{\mathcal{H}} \langle h_n, h_k \rangle_{\mathcal{H}}\big]$$

$$- \sum_{n,k\geq 0}^{N} \mathbf{E}\left[\langle U, h_k\rangle_{\mathcal{H}}\left\langle \nabla^{(2)}U, h_n \otimes h_n \otimes h_k\right\rangle_{\mathcal{H}^{\otimes(3)}}\right]$$

$$= A_1 + A_2 - A_3.$$

On the one hand, Parseval equality yields

$$A_2 = \sum_{n,k\geq 0} \mathbf{E}\left[\langle U, h_n\rangle_{\mathcal{H}} \langle U, h_k\rangle_{\mathcal{H}} \langle h_n, h_k\rangle_{\mathcal{H}}\right]$$

$$= \sum_{n\geq 0} \mathbf{E}\left[\langle U, h_n\rangle_{\mathcal{H}}^2\right] = \mathbf{E}\left[\|U\|_{\mathcal{H}}^2\right].$$

Apply once more the integration by parts formula in A_1:

$$A_1 = \sum_{n,k\geq 0}^{N} \mathbf{E}\left[\langle \nabla U, h_k \otimes h_n\rangle_{\mathcal{H}\otimes\mathcal{H}} \langle \nabla U, h_n \otimes h_k\rangle_{\mathcal{H}\otimes\mathcal{H}}\right]$$

$$+ \sum_{n,k\geq 0}^{N} \mathbf{E}\left[\langle U, h_k\rangle_{\mathcal{H}}\left\langle \nabla^{(2)}U, h_n \otimes h_k \otimes h_n\right\rangle_{\mathcal{H}^{\otimes(3)}}\right]$$

$$= \mathrm{trace}(\nabla U_N \circ \nabla U_N) + A_3,$$

since $\nabla^{(2)}$ is a symmetric operator, cf. Lemma 2.5.

Step 3 In brief, we have proved so far that

$$\mathbf{E}\left[\delta U_N^2\right] = \|U_N\|_{L^2\left(\mathbf{W}\to\mathcal{H};\,\mu\right)} + \mathbf{E}\left[\mathrm{trace}(\nabla U_N \circ \nabla U_N)\right].$$

Then, Eq. (1.24) entails that

$$\mathbf{E}\left[\delta(U_N - U_K)^2\right] \leq \|U_N - U_K\|_{L^2\left(\mathbf{W}\to\mathcal{H};\,\mu\right)}^2 + \|\nabla U_N - \nabla U_K\|_{L^2\left(\mathbf{W}\to\mathcal{H}\otimes\mathcal{H};\,\mu\right)}^2.$$

Thus, the sequence $(\delta U_N,\, N \geq 0)$ is Cauchy in $L^2(\mathbf{W} \to \mathbf{R};\, \mu)$ hence convergent toward a limit temporarily denoted by $\eta \in L^2(\mathbf{W} \to \mathbf{R};\, \mu)$. For $F \in \mathbb{D}_{1,2}$,

$$\mathbf{E}\left[\langle \nabla F, U\rangle_{\mathcal{H}}\right] = \lim_{N\to\infty} \mathbf{E}\left[\langle \nabla F, U_N\rangle_{\mathcal{H}}\right] = \lim_{N\to\infty} \mathbf{E}\left[F\,\delta U_N\right] = \mathbf{E}\left[F\,\eta\right].$$

By the very definition of the divergence, this means that $U \in \mathrm{Dom}_2\,\delta$ and $\delta U = \eta = \lim_{N\to\infty} \delta U_N$.

$\square$

During the proof, we have obtained a generalization of (2.22):

Corollary 2.4 *For $U \in \mathbb{D}_{1,2}(\mathcal{H})$, we have*

$$\nabla \delta U = U + \delta \nabla U.$$

Proof Combine (2.35) and (2.27).   □

Banach Spaces

Dual Spaces

Let X be a Banach space, i.e., a vector space with a norm for which it is complete. Its topological dual, denoted by X^*, is the set of continuous linear forms on X; i.e., the linear maps ϕ from X into **R** such that

$$\|\phi\|_{X^*} := \sup_{\|x\|_X \leq 1} |\phi(x)| < \infty.$$

To keep track of the spaces to which every term belongs to, it is often denoted

$$\phi(x) = \langle \phi, \, x \rangle_{X^*, X}. \tag{2.36}$$

A consequence of the Hahn–Banach theorem gives a very interesting way to compute the norm of an element of the original Banach space X via computations on X^*

$$\|x\|_X = \sup_{\substack{\phi \in X^* \\ \|\phi\|_{X^*} \leq 1}} |\langle \phi, \, x \rangle_{X^*, X}| \tag{2.37}$$

A Banach space is said to be reflexive whenever $X^{**} = X$. For instance, if ρ is a σ-finite measure on a space $(E, \mathcal{E})$,

- For $p \geq 1$, the dual of $L^p(E \to \mathbf{R}; \, \rho)$ is identified to the space $L^q(E \to \mathbf{R}; \, \rho)$ where $1/p + 1/q = 1$.
- The dual of $L^\infty(E \to \mathbf{R}; \, \rho)$ is strictly larger than $L^1(E \to \mathbf{R}; \, \rho)$. Hence, $L^1(E \to \mathbf{R}; \, \rho)$ is not reflexive.
- When X is a Hilbert space, the dual is isometrically isomorphic to X. Let $\kappa : X^* \to X$ be this isomorphism. Then,

$$\langle \phi, \, x \rangle_{X^*, X} = \langle \kappa(\phi), x \rangle_{X, X},$$

i.e., the duality bracket is actually the scalar product on X, which gives another reason to the notation (2.36).

- The dual space of $C([0, 1], \mathbf{R})$ is the set of finite measures on $[0, 1]$. In particular, the Dirac mass

$$\varepsilon_a : C([0, 1], \mathbf{R}) \longrightarrow \mathbf{R}$$

$$f \longmapsto f(a)$$

belongs to $(C([0, 1], \mathbf{R}))^*$.
- For $1/p < \eta$, $W_{\eta, p} \subset C([0, 1], \mathbf{R})$, hence ε_a also belongs to the dual of this space.

Dunford–Pettis Integral

It is easy to define

$$\int_a^b f(s) \mathrm{d}s$$

when f takes its value in $\mathbf{R}^d$ by simply considering this integral as the d-dimensional vector whose components are

$$\int_a^b f_i(s) \mathrm{d}s$$

for $i = 1, \cdots, d$. If f takes its value in a functional space, i.e., $f(s, .)$ is a function for any s, we may want to define the integral of f as the function

$$x \longmapsto \int_a^b f(s, x) \mathrm{d}s, \tag{2.38}$$

i.e., we integrate with respect to s for each x fixed. This will automatically raise some measurability questions. The framework of Dunford–Pettis integral is here to give a clean definition of (2.38).

Definition 2.10 A function $f : (E, \rho) \longrightarrow X$, where X is a Banach space, is weakly measurable if for any $x \in X^*$, the real-valued function $\langle x, f \rangle_{X^*, X}$ is measurable.

The same function is said to be Dunford integrable if for any $x \in X^*$, the real-valued function $\langle x, f \rangle_{X^*, X}$ belongs to $L^1(E \to \mathbf{R}; \rho)$.

Theorem 2.9 *If f is Dunford integrable, then the map*

$$X^* \longrightarrow \mathbf{R}$$

$$x \longmapsto \int_E \langle x, f \rangle \, d\rho$$

is continuous, hence belongs to X^{**}.

As far as we are concerned we will have to consider functions that are Hilbert valued, hence $X^{**} = X$ and the integral is an element of the initial space. That means there exists an element of X Hilbert denoted by $\int f d\mu$ such that

$$\left\langle \int_E f d\mu, x \right\rangle_X = \int_E \langle f, x \rangle_X \, d\mu.$$

The stochastic integral of a Hilbert valued adapted process is defined as usual. A X-valued process is said to be progressive if it is of the form

$$X(t) = \sum_{j=1}^{n} A_i \mathbf{1}_{(t_i, t_{i+1}]}(t) \, x_i$$

where $x_1, \cdots, x_n$ is a family of elements of X and A_i is $\mathcal{F}_{t_i}$-measurable. Then,

$$\int_0^t X(s) dB(s) = \sum_{j=1}^{n} A_i \left(B(t_{i+1} \wedge t) - B(t_i \wedge t) \right) \otimes x_i.$$

It is a martingale and we can then extend the notion of stochastic integral to adapted, X-valued, and square integrable processes. This yields a martingale that satisfies the Doob inequality:

$$\mathbf{E} \left[\sup_{t \leq 1} \| \int_0^t X(s) dB(s) \|_X^2 \right] \leq 4\mathbf{E} \left[\int_0^1 \| X(s) \|_X^2 ds \right].$$

Tensor Products of Banach Spaces

If we have two real-valued functions f and g defined on respective space E and F, the tensor product of f and g is simply the function of two variables $(s, t) \mapsto f(s)g(t)$, which is defined on $E \times F$. Evidently, we must take care of the topology we put on the space of such functions. There are numerous possibilities, giving raise to non-equivalent topologies. We stick here to the simplest one, namely the projective topology since it is perfectly adequate to what we have in mind. In a general setting, the tensor product of Banach spaces is defined even for Banach spaces that are not set of functions, hence the strange circumlocution via the dual spaces. We will see in the end that this amounts to the previous description when we deal with L^p-spaces.

Definition 2.11 Let X and Y be two Banach spaces, with respective dual X^* and Y^*. For $x \in X$ and $y \in Y$, $x \otimes y$ is the bilinear form defined by

$$x \otimes y : X^* \times Y^* \longrightarrow \mathbf{R}$$
$$(\eta, \zeta) \longmapsto \langle \eta, x \rangle_{X^*, X} \langle \zeta, y \rangle_{Y^*, Y}. \tag{2.39}$$

We now define the topology on the space spanned by the $x \otimes y$.

Definition 2.12 (See [6, chapter 2]) The *projective* tensor product of X and Y, denoted by $X \otimes Y$, is the completion of the vector space spanned by the finite linear combinations of some $x \otimes y$ for $x \in X$ and $y \in Y$, equipped with the norm

$$\|z\|_{X \otimes Y} = \inf \left\{ \sum_{i=1}^{n} \|x_i\|_X \|y_i\|_Y, \ z = \sum_{i=1}^{n} x_i \otimes y_i \right\}. \tag{2.40}$$

Example (Tensor Product of L^2 Spaces) If $X = L^2(E \to \mathbf{R}; \ m)$ and $Y = L^2(F \to \mathbf{R}; \ \rho)$, then $X^* \simeq X$ and $Y^* \simeq Y$. Furthermore,

$$\langle x \otimes y, \ f \otimes g \rangle_{X \otimes Y, \ X^* \otimes Y^*} = \int_E x(s) f(s) \mathrm{d}m(s) \int_F y(t) g(t) \mathrm{d}\rho(t)$$
$$= \iint_{E \times F} x(s) y(t) \ f(s) g(t) \mathrm{d}m(s) \ \mathrm{d}\rho(t).$$

Thus, $x \otimes y$ can be identified with the function of two variables $(s, t) \mapsto x(s) y(t)$ and as we shall see below in a more general case

$$L^2(E \to \mathbf{R}; \ m) \otimes L^2(F \to \mathbf{R}; \ \rho) \simeq L^2(E \times F \to \mathbf{R}; \ m \otimes \rho).$$

In the definition of the norm on $X \otimes Y$, we need to take the infimum of all the possible representations of z as a linear combinations of elementary tensor products since such a representation is by no means unique.

Example (Decomposition of an L as a Sum of Two Rectangles) One of the simplest situation we can imagine is the tensor product of $L^1(\mathbf{R} \to \mathbf{R}; \ \ell)$ by itself. The function

$$\mathbf{1}_{[0,1]}(s) \otimes \mathbf{1}_{[0,2]}(t) + \mathbf{1}_{[1,2]}(s) \otimes \mathbf{1}_{[1,2]}(t)$$

can be equally written as

$$\mathbf{1}_{[0,1]}(s) \otimes \mathbf{1}_{[0,1]}(t) + \mathbf{1}_{[0,2]}(s) \otimes \mathbf{1}_{[1,2]}(t).$$

We then see that an element of $\mathrm{span}\{x \otimes y, \ x \in X, y \in Y\}$ may have several representations, thus the need to take the infimum in (2.40).

Proposition 2.1 *For X and Y two reflexive Banach spaces, i.e., $(X^*)^* = X$ and $(Y^*)^* = Y$. The dual of $W = X \otimes Y$ is the space $W^* = X^* \otimes Y^*$ with the duality pairing:*

$$\langle w^*, w \rangle_{W*,W} = \sum_{i,j} \langle x_i^*, x_j \rangle_{X*,X} \, \langle y_i^*, y_j \rangle_{Y*,Y}$$

where $w = \sum_j x_j \otimes y_j \in X \otimes Y$ and $w^* = \sum_i x_i^* \otimes y_i^* \in X^* \otimes Y^*$. Moreover,

$$\|w^*\|_{W^*} := \sup_{\|w\|_W = 1} |\langle w^*, w \rangle_{W*,W}|$$

$$= \sup\Big\{ |\langle w^*, x \otimes y \rangle_{W*,W}|, \ \|x\|_X = 1, \ \|y\|_Y = 1 \Big\}. \qquad (2.41)$$

This proposition is important as it says that to compute the norm of an element of W^*, it is sufficient to test it against *simple* tensor products.

Let X be a Banach space and ν a measure on a space E. The set $L^p(E \to X; \nu)$ is the space of functions ψ from E into X such that

$$\int_E \|\psi(x)\|_X^p d\nu(x) < \infty.$$

Theorem 2.10 ([6, p. 30]) *For X a Banach space, the space $L^p(E \to \mathbf{R}; \nu) \otimes X$ is isomorphic to $L^p(E \to X; \nu)$.*

Moreover, if $X = L^p(F \to \mathbf{R}; \rho)$, then $L^p(E \to X; \nu)$ is isometrically isomorphic to $L^p(E \times F \to \mathbf{R}; \nu \otimes \rho)$. Moreover, the set of simple functions, i.e., functions of the form

$$\sum_{j=1}^{n} f_j(s) \psi_j(x)$$

where $f_j \in L^p(E \to \mathbf{R}; \nu)$ and $\psi_j \in L^p(F \to \mathbf{R}; \rho)$, is dense into $L^p(E \times F \to \mathbf{R}; \nu \otimes \rho)$.

Convergence, Strong, and Weak

Definition 2.13 (Weak Convergence) A sequence $(x_n, n \geq 0)$ is said to be weakly convergent in a Banach space X, if for every $\eta \in X^*$, $(\langle \eta, x_n \rangle_{X*,X}, n \geq 0)$ is convergent.

Remark 2.3 Since $|\langle \eta, x_n - x \rangle_{X*,X}| \leq \|\eta\|_{X^*}\|x_n - x\|_X$, strong convergence implies weak convergence but the converse is false. For instance, let $(e_n, n \geq$

0) be a complete orthonormal basis in a Hilbert space X, on the one hand $\|e_n\|_X = 1$. On the other hand, according to Parseval equality, for $\eta \in X^* = X$, $\|\eta\|_X^2 = \sum_n |\langle \eta, e_n \rangle_X|^2$. Hence, $(\langle \eta, x_n \rangle_{X^*, X}, n \geq 0)$ converges weakly to 0. The convergence cannot hold in the strong sense.

Proposition 2.2 (Eberlein-Shmulyan [9, p. 141]) *Let X be a reflexive Banach space, i.e., $(X^*)^* = X$. Then, any strongly bounded sequence admits a weakly convergent subsequence.*

Remark 2.4 For any measure, L^p spaces are reflexive only for $p \neq 1, \infty$. We do have that the dual of L^1 is L^∞, but the dual of L^∞ is larger than L^1.

Proposition 2.3 (Mazur [9, p. 120]) *Let $(x_n, n \geq 0)$ be a weakly convergent subsequence in a Banach space X and set x its limit. Then, for any $\epsilon > 0$, there exist n and $(\alpha_i, 1 \leq i \leq n)$ such that $\alpha_i \geq 0$, $\sum_i \alpha_i = 1$ and*

$$\left\| \sum_{i=1}^n \alpha_i x_{n_i} - x \right\|_X \leq \epsilon.$$

2.3 Problems

2.1 Let $F \in \mathbb{D}_{p,1}$ and $\epsilon > 0$. Set $\phi_\epsilon(x) = \sqrt{x^2 + \epsilon^2}$.

1. Show that $\phi_\epsilon(F) \in \mathbb{D}_{p,1}$.
2. Show that $|F| \in \mathbb{D}_{p,1}$ and that

$$\dot{\nabla}_s |F| = \begin{cases} \dot{\nabla}_s F & \text{if } F > 0 \\ 0 & \text{if } F = 0 \\ -\dot{\nabla}_s F & \text{if } F < 0. \end{cases}$$

3. If $G \in \mathbb{D}_{p,1}$, compute $\nabla(F \vee G)$.

   Let B be the standard Brownian motion on $[0, 1]$ and $M = \sup_{t \in [0,1]} B(s)$. Let $\mathbb{Q} \cap [0, 1] = \{t_n, n \geq 0\}$. Consider

$$M_n = \sup_{s \in \{t_1, \cdots, t_n\}} B(s).$$

We admit that B attains its maximum at a unique point T, almost-surely. Let

$$T = \arg\max_{s \in [0,1]} B(s).$$

4. Show that M_n belongs to $\mathbb{D}_{p,1}$ and compute $\dot{\nabla} M_n$.
5. Prove that $M \in \mathbb{D}_{p,1}$ and that $\dot{\nabla} M = \mathbf{1}_{[0,T]}$.

2.2 (Iterated Divergence) For $U \in \mathcal{S}(\mathcal{H})$, i.e.,

$$U = \sum_{j=1}^{n} f_j(\delta h_1, \cdots, \delta h_m) v_j$$

where $(v_1, \cdots, v_n)$ belong to $\mathcal{H}$ and f_j in the Schwartz space on $\mathbf{R}^m$. Let $\delta^{(2)}$ defined by the duality

$$\mathbf{E}\left[\delta^2 u^{\otimes(2)} G\right] = \mathbf{E}\left[\left\langle u^{\otimes(2)}, \nabla^{(2)} G\right\rangle_{\mathcal{H} \otimes \mathcal{H}}\right]$$

for any $G \in \mathbb{D}_{2,2}$. Show that

$$\delta^2(U^{\otimes(2)}) = (\delta U)^2 - \|U\|_{\mathcal{H}}^2 - \text{trace}(\nabla U \circ \nabla U) - 2\delta(\langle \nabla U, U \rangle_{\mathcal{H}}).$$

2.3 (Stratonovich Integral) The Itô integral has a major drawback: Its differential is not given by the usual formula but by the Itô formula. On the other hand, the Stratonovich integral does satisfy the usual rule of differentiation but does not yield a martingale! We see in this problem that the Stratonovich integral can be computed with δ and ∇. For $T_n = \{0 = t_0 < t_1 = 1/n < \ldots < t_n = 1\}$, let

$$\mathrm{d}B_{T_n}(t) = \sum_{i=0}^{n-1} \frac{B(t_{i+1}) - B(t_i)}{t_{i+1} - t_i} \mathbf{1}_{[t_i, t_{i+1}]}(t)\mathrm{d}t$$

and

$$B_{T_n}(t) = \int_0^t \mathrm{d}B_{T_n}(s) = \sum_{i=0}^{n-1} \left(B(t_i) + \frac{B(t_{i+1}) - B(t_i)}{t_{i+1} - t_i}(t - t_i)\right) \mathbf{1}_{[t_i, t_{i+1}]}(t)$$

be the linear affine interpolation of B. For any $\mathcal{H}$-valued random variable U, consider the Riemann-like sum

$$S_{T_n}^U = \int_0^1 \dot{U}(s)\mathrm{d}B_{T_n}(s) = \sum_{i=0}^{n-1} \frac{B(t_{i+1}) - B(t_i)}{t_{i+1} - t_i} \int_{t_i}^{t_{i+1}} \dot{U}(t)\mathrm{d}t.$$

The process U is said to be Stratonovich integrable if the sequence $(S_{T_n}^U, n \geq 0)$ converges in probability as n goes to infinity.

Assume that U belongs to $\mathbb{D}_{1,2}(\mathcal{H})$ and that the map

$$[0, 1] \times [0, 1] \longrightarrow \mathbf{R}$$

$$(s, t) \longmapsto \dot{\nabla}_s \dot{U}(t)$$

is continuous.

1. Show that U is Stratonovich integrable and

$$\lim_{n\to\infty} S_{T_n}^U = \delta U + \int_0^1 \dot{\nabla}_r \dot{U}(r)\mathrm{d}r.$$

Indication: Verify that

$$S_{T_n}^U = \sum_{i=0}^{n-1} \frac{1}{t_{i+1} - t_i} \delta(I^1(\mathbf{1}_{[t_i,t_{i+1}]})) \int_{t_i}^{t_{i+1}} \dot{U}(t)\mathrm{d}t.$$

Apply (2.27).
2. Find

$$\lim_{n\to\infty} \sum_{i=0}^{n-1} \frac{1}{2}\left(\dot{U}(t_i) + \dot{U}(t_{i+1})\right)(B(t_{i+1}) - B(t_i)).$$

2.4   Notes and Comments

The presentation of the so-called Gross–Sobolev gradient avoids deliberately chaos decomposition as in [7, 8]. It requires to invoke sophisticated theorems from functional analysis, but the reward will be apparent in the chapter about fractional Brownian motion. For other approaches, see [4, 5]. The definition of the gradient without cylindric functions has been investigated in [1, 2, 8].

References

1. S. Kusuoka, The nonlinear transformation of Gaussian measure on Banach space and absolute continuity. I. J. Fac. Sci. Univ. Tokyo Sect. IA Math. **29**(3), 567–597 (1982)
2. S. Kusuoka, The nonlinear transformation of Gaussian measure on Banach space and its absolute continuity. II. J. Fac. Sci. Univ. Tokyo Sect. IA Math. **30**(1), 199–220 (1983)
3. A. Lejay, Yet another introduction to rough paths, in *Séminaire de Probabilités XLII* (Springer, Berlin, 2009), pp. 1–101
4. D. Nualart, *The Malliavin Calculus and Related Topics* (Springer, Berlin, 1995)
5. N. Privault, *Stochastic Analysis in Discrete and Continuous Settings with Normal Martingales*, vol. 1982. Lecture Notes in Mathematics (Springer, Berlin, 2009)
6. R.A. Ryan, *Introduction to Tensor Products of Banach Spaces* (Springer, London, 2002)
7. A.S. Üstünel, *An Introduction to Analysis on Wiener Space*, vol. 1610. Lectures Notes in Mathematics (Springer, Berlin, 1995)
8. A.S. Üstünel, M. Zakai, *Transformation of Measure on Wiener Space* (Springer, Berlin, 2000)
9. K. Yosida, *Functional Analysis* (Springer, Berlin, 1995)

Chapter 3
Wiener Chaos

Abstract Chaos are the eigenspaces of the $L = -\delta\nabla$ they play a major rôle in the Hilbertian on the Wiener space as ∇ and δ have simple expressions on chaos elements. They can also be constructed as iterated integrals with respect to the Brownian motion and as such replace the orthonormal polynomials in the usual deterministic calculus.

The next results will mainly be obtained by density reasoning. That means, the desired formula is established for a small but dense subset of functionals. It is then generalized to a wider set of functionals by passing to the limit. A small but rich enough set is the set of Doléans-Dade exponentials: For $h \in \mathcal{H}$,

$$\Lambda_h = \exp\left(\delta h - \frac{1}{2}\|h\|_H^2\right).$$

Lemma 3.1 (Density of Doléans-Dade Exponentials) *The set of Doléans-Dade exponentials:*

$$\mathcal{E} = \operatorname{span}\{\Lambda_h,\ h \in \mathcal{H}\}$$

is dense in $L^2(W \to \mathbf{R};\ \mu)$.

Proof Let $Z \in L^2(W \to \mathbf{R};\ \mu)$ be orthogonal to all the elements of $\mathcal{E}$. Let $t_0 = 0 < t_1 \ldots t_n \le 1$ and $(z_1, \cdots, z_n) \in \mathbf{C}^n$, for

$$h = \sum_{j=1}^{n} z_j\,(t_j \wedge . - t_{j-1} \wedge .),$$

we have

$$\Lambda_h = \exp\left(\sum_{j=1}^{n} z_j\Big(B(t_j) - B(t_{j-1})\Big) - \frac{1}{2}\sum_{j=1}^{n} z_j^2(t_j - t_{j-1})\right).$$

L. Decreusefond, *Selected Topics in Malliavin Calculus*,
Bocconi & Springer Series 10, https://doi.org/10.1007/978-3-031-01311-9_3

Consider the map

$$\mathfrak{G} : \mathbf{C}^n \longrightarrow \mathbf{C}$$

$$z = (z_1, \cdots, z_n) \longmapsto \mathbf{E}\big[Z \exp(\delta h)\big].$$

The hypothesis says that $\mathfrak{G}$ is null on $\mathbf{R}^n$. We now prove that $\mathfrak{G}$ is holomorphic on $\mathbf{C}^n$, hence null everywhere. For any $j \in \{1, \cdots, n\}$, we can expand the exponential into a series with respect to z_j,

$$\exp\big(z_j\big(B(t_j) - B(t_{j-1})\big)\big) = \sum_{k=0}^{\infty} \frac{1}{k!}\big(B(t_j) - B(t_{j-1})\big)^k z_j^k.$$

Since Gaussian random variables have finite moments of every order, multiple applications of Hölder inequality entail that $\mathfrak{G}$ has a series expansion valid on $\mathbf{C}^n$, hence is holomorphic.

It follows that $\mathfrak{G}$ is null on $(i\mathbf{R})^n$, i.e.,

$$\mathbf{E}\left[Z \exp\left(i \sum_{j=1}^{n} \alpha_j\big(B(t_j) - B(t_{j-1})\big) - \frac{1}{2}\sum_{j=1}^{n}\alpha_j^2(t_j - t_{j-1})\right)\right] = 0,$$

for any $\alpha = (\alpha_1, \cdots, \alpha_n) \in \mathbf{R}^n$. If $\phi \in$ Schwartz$(\mathbf{R}^n)$, let $\hat{\phi}$ denote its Fourier transform, we have

$$\mathbf{E}\,[Z\,\phi(B(t_1), B(t_2) - B(t_1), \cdots)]$$

$$= \int_{\mathbf{R}^n} \hat{\phi}(\alpha_1, \cdots, \alpha_n)\mathbf{E}\,[Z\mathfrak{G}(i\alpha)]\,\mathrm{d}\alpha_1 \ldots \mathrm{d}\alpha_n = 0.$$

This means that Z is orthogonal to cylindrical functions that are known to be dense in $L^2(W \to \mathbf{R};\ \mu)$, hence Z is null.   □

3.1   Chaos Decomposition

In classical analysis, polynomials are interesting because their derivative is easy to compute and the vector space they span is often dense. The first feature comes from the identity

$$\frac{t^n}{n!} = \int_0^t \int_0^{t_1} \cdots \int_0^{t_{n-1}} \mathrm{d}t_n \ldots \mathrm{d}t_1.$$

With this presentation, it is straightforward that

$$\left(\frac{t^n}{n!}\right)' = \frac{t^{n-1}}{(n-1)!}$$

so that we can hope for a similar behavior if we find the good generalization. It is natural to consider iterated integrals with respect to the Brownian motion:

$$\int_0^1 \int_0^{t_1} \cdots \int_0^{t_{n-1}} dB(t_n) \ldots dB(t_1). \tag{3.1}$$

The analog of putting a coefficient in front of the monomial t^n is here to integrate a deterministic function f of n variables as in (3.2). We so define the Wiener chaos (of order n if there are n variables), which turn to be the analog of the Hermite polynomials in $\mathbf{R}^n$.

Definition 3.1 (Iterated Integrals on a Simplex) For $t \in (0, 1]$, let

$$\mathcal{T}_n(t) = \left\{ (t_1, \cdots, t_n) \in [0, 1]^n, \ 0 \le t_1 < \ldots < t_n \le t \right\}.$$

For $f \in L^2\big(\mathcal{T}_n(t) \to \mathbf{R}; \ell\big)$, set

$$J_n(f)(t) = \int_0^t dB(t_n) \int_0^{t_n} dB(t_{n-1}) \ldots \int_0^{t_2} f(t_1, \cdots, t_n) dB(t_1), \tag{3.2}$$

where the integrals are Itô integrals. For the sake of notations, set $\mathcal{T}_n = \mathcal{T}_n(1)$ and $J_n(f) = J_n(f)(1)$.

For $n = 0$, $\mathcal{T}_0$ is reduced to one point and elements of $L^2\big(\mathcal{T}_0(t) \to \mathbf{R}; \ell\big)$ are simply constant functions. Furthermore, $J_0(a) = a$.

The structure of $\mathcal{T}_n(t)$ ensures that at each internal integral, the integrand is adapted. Moreover,

$$J_n(f)(t) = \int_0^t J_{n-1}(f(., t_n))(t_n) dB(t_n). \tag{3.3}$$

The Itô isometry then entails that

Theorem 3.1 *We have*

$$\mathbf{E}\left[J_n(f)J_m(g)\right] = \begin{cases} 0 & \text{if } n \ne m \\ \int_{\mathcal{T}_n} fg d\ell & \text{if } n = m. \end{cases} \tag{3.4}$$

Proof For $n = 0$, Eq. (3.3) entails that

$$\mathbf{E}[1.J_m(f)] = \mathbf{E}[J_m(f)] = 0 \tag{3.5}$$

for any f.

For $n = 1$ and $m > 1$, the Itô isometry formula states that

$$\mathbf{E}[J_1(f)J_m(g)] = \mathbf{E}\left[\int_0^1 f(s)dB(s)\int_0^1 J_{m-1}(g(.,s))dB(s)\right]$$

$$= \mathbf{E}\left[\int_0^1 f(s)J_{m-1}(g(.,s))ds\right]$$

$$= \int_0^1 f(s)\,\mathbf{E}[J_{m-1}(g(.,s))]\,ds = 0$$

in view of the induction hypothesis for $n = 0$.

For $1 < n \leq m$, a repeated application of the Itô isometry formula yields

$$\mathbf{E}[J_n(f)J_m(g)] = \mathbf{E}\left[\int_{\mathcal{T}_n} f(t_1,\cdots,t_n)J_{m-n}\Big(g(.,t_1,\cdots,t_n)\Big)dt_1\ldots dt_n\right].$$

In view of (3.5), this quantity is null if $n - m \neq 0$ and is clearly equal to $\int_{\mathcal{T}} fg d\ell$ if $n = m$. $\qquad\square$

We wish to extend this notion of iterated integral to function defined on the whole cube $[0, 1]^n$, but we cannot get rid of the adaptability condition. It is then crucial to remark that for $f : [0, 1]^n \to \mathbf{R}$ symmetric,

$$\int_{[0,1]^n} f d\ell = n! \int_{\mathcal{T}_n} f d\ell,$$

since for any permutation σ of $\{1, \cdots, n\}$, the integral of f on $\mathcal{T}_n$ is equal to its integral on

$$\sigma\mathcal{T}_n = \left\{(t_1,\cdots,t_n) \in [0, 1]^n,\ 0 \leq t_{\sigma(1)} < \ldots < t_{\sigma(n)} \leq 1\right\}.$$

This motivates the following definition of the iterated integral:

Definition 3.2 (Generalized Iterated Integrals) Let $L_s^2 = L_s^2([0, 1]^n \to \mathbf{R};\ \ell)$ be the set of symmetric functions on $[0, 1]^n$, square integrable with respect to the Lebesgue measure. For $f \in L_s^2$,

$$J_n^s(f) = n! J_n(f 1_{\mathcal{T}_n}).$$

If f belongs to $L^2([0, 1]^n \to \mathbf{R};\ \ell)$ but is not necessarily symmetric,

$$J_n^s(f) = J_n^s(f^s),$$

where f^s is the symmetrization of f:

$$f^s(t_1, \cdots, t_n) = \frac{1}{n!} \sum_{\sigma \in \mathfrak{S}_n} f(t_{\sigma(1)}, \cdots, t_{\sigma(n)}).$$

In view of Eq. (3.4), for $f, g \in L_s^2$, we have

$$\mathbf{E}\left[J_n^s(f) J_m^s(g)\right] = \begin{cases} 0 & \text{if } n \neq m \\ (n!)^2 \displaystyle\int_{\mathcal{T}_n} fg \, d\ell = n! \int_{[0,1]^n} fg \, d\ell & \text{if } n = m. \end{cases} \tag{3.6}$$

Doléans-Dade Exponentials Behave as Usual Exponentials
The Doléans-Dade exponentials have *mutatis mutandis* the same series expansion as the usual exponential has.

Theorem 3.2 (Chaos Expansion of Doléans-Dade Exponentials) *Let h belong to $\mathcal{H}$. Then,*

$$\Lambda_h = 1 + \sum_{n=1}^{\infty} J_n(\dot{h}^{\otimes n} \mathbf{1}_{\mathcal{T}_n}) = 1 + \sum_{n=1}^{\infty} \frac{1}{n!} J_n^s(\dot{h}^{\otimes n}), \tag{3.7}$$

where the convergence holds in $L^2(\mathbf{W} \to \mathbf{R}; \mu)$.

Proof **Step 1** Let

$$\Lambda_h(t) = \exp\left(\int_0^t \dot{h}(s) \, dB(s) - \frac{1}{2} \int_0^1 \dot{h}(s)^2 \, ds\right).$$

The Itô calculus says that

$$\Lambda_h(t) = 1 + \int_0^t \Lambda_h(s) \, \dot{h}(s) \, dB(s),$$

hence

$$\begin{aligned}
\Lambda_h(t) &= 1 + \int_0^t \Lambda_h(s) \, \dot{h}(s) \, dB(s) \\
&= 1 + \int_0^t \left(1 + \int_0^s \Lambda_h(r) \dot{h}(r) \, dB(r)\right) \dot{h}(s) \, dB(s) \\
&= 1 + \int_0^t \dot{h}(s) \, dB(s) + \int_0^t \left(\int_0^s \Lambda_h(r) \dot{h}(s) \dot{h}(r) \, dB(r)\right) dB(s)
\end{aligned}$$

$$= 1 + \sum_{k=1}^{n} J_k(\dot{h}^{\otimes k} \mathbf{1}_{\mathcal{T}_k}) + \int_{\mathcal{T}_n} \prod_{j=1}^{n} \dot{h}(s_j) \, \Lambda_h(s_1) \mathrm{d}B(s_1) \ldots \mathrm{d}B(s_n)$$

$$= 1 + \sum_{k=1}^{n} J_k(\dot{h}^{\otimes k} \mathbf{1}_{\mathcal{T}_k}) + R_n.$$

Step 2 It thus remains to show that R_n tends to 0 as n goes to infinity. According to (3.4),

$$\mathbf{E}\left[R_n^2\right] = \int_{\mathcal{T}_n} \prod_{j=1}^{n} \dot{h}(s_j)^2 \mathbf{E}\left[\Lambda_h(s_n)^2\right] \mathrm{d}s_1 \ldots \mathrm{d}s_n. \tag{3.8}$$

Moreover,

$$\mathbf{E}\left[\Lambda_h(s)^2\right] = \mathbf{E}\left[\exp\left(2\int_0^s h(u)\mathrm{d}B(u) - \int_0^s h^2(u)\mathrm{d}u\right)\right]$$

$$= \mathbf{E}\left[\Lambda_{2h}(s)\right] \exp(\|h\|_{\mathcal{H}}^2)$$

$$= \exp(\|h\|_{\mathcal{H}}^2).$$

Plug this new expression into Eq. (3.8) to obtain

$$\mathbf{E}\left[R_n^2\right] = \exp(\|h\|_{\mathcal{H}}^2) \int_{\mathcal{T}_n} \prod_{j=1}^{n} \dot{h}(s_j)^2 \mathrm{d}s_1 \ldots \mathrm{d}s_n$$

$$= \exp(\|h\|_{\mathcal{H}}^2) \frac{1}{n!} \int_{[0,1]^n} \prod_{j=1}^{n} \dot{h}(s_j)^2 \, \mathrm{d}s_1 \ldots \mathrm{d}s_n$$

$$= \exp(\|h\|_{\mathcal{H}}^2) \frac{1}{n!} \prod_{j=1}^{n} \int_{[0,1]} \dot{h}(s_j)^2 \, \mathrm{d}s_j$$

$$= \exp(\|h\|_{\mathcal{H}}^2) \frac{1}{n!} \|h\|_{\mathcal{H}}^{2n} \xrightarrow{n \to \infty} 0.$$

The result follows.

$\square$

The Fock Space Plays the Rôle of R[X]

When dealing with polynomials of arbitrary degree, we need to consider $\mathbf{R}[X] = \bigcup_{k=0}^{\infty} \mathbf{R}_k[X]$. The equivalent structure is the Fock space where the monomial X is replaced by a function h of $\mathcal{H}$ and X^n by the tensor product $h^{\otimes(n)}$.

Definition 3.3 (Fock Space) The Fock space $\mathfrak{F}_\mu(\mathcal{H})$ is the completion of the direct sum of the tensor powers of $\mathcal{H}$:

$$\mathfrak{F}_\mu(\mathcal{H}) = \mathbf{R} \oplus \bigoplus_{n=1}^\infty \mathcal{H}^{\otimes n}.$$

It is a Hilbert space when equipped with the norm

$$\left\| \oplus_{n=0}^\infty h_n \right\|_{\mathfrak{F}_\mu(\mathcal{H})}^2 = \sum_{n=0}^\infty \frac{1}{n!} \| h_n \|_{\mathcal{H}^{\otimes n}}^2.$$

If we want to generalize the chaos decomposition of Doléans-Dade exponentials to any random variables on W, we first need to express the right-hand side of (3.7) in an intrinsic way. Remark that for $F = \Lambda_h$,

$$\nabla^{(n)} F = F h^{\otimes n}, \text{ hence } \mathbf{E}\left[\nabla^{(n)} F \right] = h^{\otimes n},$$

so that we have

$$F = \mathbf{E}[F] + \sum_{n=1}^\infty \frac{1}{n!} J_n^s \big(\mathbf{E}\big[\widetilde{\nabla^{(n)} F} \big] \big). \tag{3.9}$$

By linearity, the same holds true for any $F \in \mathcal{E}$. If we want to pass to the limit, we must prove that each term in the expansion (3.9) is well defined and that the application that maps a random variable F to its series expansion is continuous. The first difficulty is that we only assumed F to be square integrable.

> **Expectation Is a Smoothing Operator**
> There is no reason why F should be infinitely differentiable hence, a priori, the expression $\mathbf{E}\left[\nabla^{(n)} F \right]$ has no signification. However, it turns out that the composition of the derivation and of the expectation can be defined even when F is only square integrable.

Theorem 3.3 *The map*

$$\Upsilon : \mathcal{E} \subset L^2(W \to \mathbf{R}; \mu) \longrightarrow \mathfrak{F}_\mu(\mathcal{H})$$

$$F \longmapsto \bigoplus_{n=0}^\infty \mathbf{E}\left[\nabla^{(n)} F \right],$$

admits a continuous extension defined on $L^2(W \to \mathbf{R}; \mu)$. We denote by $\Upsilon_n F$, the n-th term of the right-hand side: $\Upsilon_n F = \mathbf{E}[\nabla^{(n)} F]$ for $F \in \mathcal{E}$.

Proof Start from (3.9), since the chaos are orthogonal, for any $F \in \mathcal{E}$,

$$
\mathbf{E}\left[F^2\right] = \sum_{n=0}^{\infty} \frac{1}{n!^2} \mathbf{E}\left[J_n^s(\mathbf{E}\widetilde{[\nabla^{(n)}F]})^2\right]
$$

$$
= \sum_{n=0}^{\infty} \frac{1}{n!} \left\| \mathbf{E}\widetilde{[\nabla^{(n)}F]} \right\|^2_{L^2([0,1]^n \to \mathbf{R};\, \ell^{\otimes n})}
$$

$$
= \sum_{n=0}^{\infty} \frac{1}{n!} \left\| \mathbf{E}[\nabla^{(n)}F] \right\|^2_{\mathcal{H}^{\otimes n}}.
$$

This is equivalent to say that

$$
\|\Upsilon F\|_{\mathfrak{F}_\mu(\mathcal{H})} = \|F\|_{L^2(W \to \mathbf{R};\, \mu)}. \tag{3.10}
$$

If $(F_n,\ n \geq 1)$ is a sequence of elements of $\mathcal{E}$, which converges to F in $L^2(W \to \mathbf{R};\ \mu)$, the sequence $(\Upsilon F_n,\ n \geq 1)$ is Cauchy in the Hilbert space $\mathfrak{F}_\mu(\mathcal{H})$, hence convergent. Then, ΥF can be unambiguously defined as $\lim_{n \to \infty} \Upsilon F_n$ and (3.10) holds for any $F \in L^2(W \to \mathbf{R};\ \mu)$. $\qquad\square$

We are now ready to state and prove the chaos decomposition. Remark that a necessary condition for a function of the real variable to have an infinite series expansion is that it is infinitely many times differentiable. For chaos decomposition, it is sufficient that F is square integrable thanks to Theorem 3.3.

Theorem 3.4 (Chaos Decomposition) *For any $F \in L^2(W \to \mathbf{R};\ \mu)$,*

$$
F = \mathbf{E}[F] + \sum_{n=1}^{\infty} \frac{1}{n!} J_n^s(\widetilde{\Upsilon_n F}). \tag{3.11}
$$

This can be formally written as

$$
F = \mathbf{E}[F] + \sum_{n=1}^{\infty} \frac{1}{n!} J_n^s(\mathbf{E}\widetilde{[\nabla^{(n)}F]}),
$$

keeping in mind that $\mathbf{E}[\nabla^{(n)}F]$ is defined through Υ for general random variables.

The chaos decomposition means that $\mathfrak{F}_\mu(\mathcal{H})$ is isometrically isomorphic to $L^2(W \to \mathbf{R};\ \mu)$.

We denote by $\mathfrak{C}_k$, the k-th chaos, i.e.,

$$
\mathfrak{C}_k = \mathrm{span}\big\{ J_n^s(f_n),\ f_n \in L_s^2([0,1]^n \to \mathbf{R},\ \ell) \big\}.
$$

Proof **Step 1** Equation (3.9) indicates that the result holds for $F \in \mathcal{E}$.

Step 2 Let $(F_k, \ k \geq 1)$ be a sequence of elements of $\mathcal{E}$ converging to F in $L^2(\mathrm{W} \to \mathbf{R}; \ \mu)$. Since Υ is continuous from $L^2(\mathrm{W} \to \mathbf{R}; \ \mu)$ into $\mathfrak{F}_\mu(\mathcal{H})$,

$$\Upsilon F_k \xrightarrow[k \to \infty]{\mathfrak{F}_\mu(\mathcal{H})} \Upsilon F.$$

Since the chaos are orthogonal in $L^2(\mathrm{W} \to \mathbf{R}; \ \mu)$

$$\mathbf{E}\left[\left|\sum_{n=1}^{\infty} \frac{1}{n!} J_n^s(\Upsilon_n F_k) - \sum_{n=1}^{\infty} \frac{1}{n!} J_n^s(\Upsilon_n F)\right|^2\right] = \sum_{n=1}^{\infty} \frac{1}{n!} \mathbf{E}\left[|\Upsilon_n F_k - \Upsilon_n F|^2\right]$$

$$= \|\Upsilon(F_k - F)\|_{\oplus_{n=0}^{\infty} \mathcal{H}^{\otimes n}}^2.$$

This means that

$$0 = F_k - \sum_{n=0}^{\infty} \frac{1}{n!} J_n^s(\Upsilon_n F_k) \xrightarrow[k \to \infty]{L^2(\mathrm{W} \to \mathbf{R}; \ \mu)} F - \sum_{n=0}^{\infty} \frac{1}{n!} J_n^s(\Upsilon_n F).$$

The proof is thus complete.

$\square$

We already know that Wiener integral and divergence of deterministic functions do coincide. We can now close the loop and show that this still holds at any order : Iterated integrals and iterated divergence do coincide.

Theorem 3.5 (Iterated Integrals and Iterated Divergence Coincide) *For any* $h \in \mathcal{H}$,

$$J_n^s(\dot{h}^{\otimes n}) = \delta^n h^{\otimes n}.$$

Hence, for any $F \in L^2(\mathrm{W} \to \mathbf{R}; \ \mu)$,

$$F = \mathbf{E}[F] + \sum_{n=1}^{\infty} \frac{1}{n!} \delta^n(\Upsilon_n F). \tag{3.12}$$

Proof **Step 1** For $F = \Lambda_k$, thanks to (1.15), we have

$$F(\omega + \tau h) = F(\omega) \exp\left(\tau \langle h, k \rangle_{\mathcal{H}}\right),$$

hence $\tau \mapsto F(\omega + \tau h)$ is analytic. Furthermore,

$$
\begin{aligned}
\left. \frac{d^n}{d\tau^n} F(\omega + \tau h) \right|_{\tau=0} &= F(\omega) \, \langle h, k \rangle_{\mathcal{H}}^n \\
&= F(\omega) \, \langle h^{\otimes n}, k^{\otimes n} \rangle_{\mathcal{H}^{\otimes n}} \\
&= \langle \nabla^{(n)} F(\omega), h^{\otimes n} \rangle_{\mathcal{H}^{\otimes n}},
\end{aligned}
$$

since $\nabla^{(n)} \Lambda_k = \Lambda_k \, k^{\otimes n}$.

Step 2 The Taylor–MacLaurin formula then says that

$$
\begin{aligned}
F(\omega + \tau h) &= F(\omega) + \sum_{n=1}^{\infty} \frac{\tau^n}{n!} \frac{d^n}{d\tau^n} F(\omega + \tau h) \Big|_{\tau=0} \\
&= F(\omega) + \sum_{n=1}^{\infty} \frac{\tau^n}{n!} \langle \nabla^{(n)} F(\omega), h^{\otimes n} \rangle_{\mathcal{H}^{\otimes n}}.
\end{aligned}
$$

Hence,

$$
\int_W F(\omega + \tau h) \, d\mu(\omega) = \mathbf{E}[F] + \sum_{n=1}^{\infty} \frac{\tau^n}{n!} \mathbf{E}\left[F \, \delta^n h^{\otimes n}\right].
$$

By linearity, this still holds for $F \in \mathcal{E}$.

Step 3 On the other hand, the Cameron–Martin theorem and Theorem 3.2 induce that for $F \in \mathcal{E}$:

$$
\begin{aligned}
\int_W F(\omega + \tau h) \, d\mu(\omega) &= \mathbf{E}[F \, \Lambda_{\tau h}] \\
&= \mathbf{E}[F] + \sum_{n=1}^{\infty} \frac{1}{n!} \mathbf{E}\left[F \, J_n^s((\tau \dot{h})^{\otimes n})\right] \\
&= \mathbf{E}[F] + \sum_{n=1}^{\infty} \frac{\tau^n}{n!} \mathbf{E}\left[F \, J_n^s(\dot{h}^{\otimes n})\right].
\end{aligned}
$$

By identification of the coefficient of the two power series, we get

$$
\mathbf{E}\left[F \, J_n^s(\dot{h}^{\otimes n})\right] = \mathbf{E}\left[F \, \delta^n h^{\otimes n}\right], \quad \forall F \in \mathcal{E}.
$$

Since $\mathcal{E}^{\perp} = \{0\}$, the result follows. $\qquad \square$

Example (Chaos Representation of B_t^2) In order to play with the notations, compute the chaos decomposition of B_t^2. First, we know that $\mathbf{E}\left[B_t^2\right] = t$. Then,

$$\dot{\nabla}_s B_t^2 = 2\, B_t \mathbf{1}_{[0,t]}(s) \text{ hence } \mathbf{E}\left[\dot{\nabla}_s B_t^2\right] = 0.$$

As to the second derivative,

$$\dot{\nabla}_{r,s}^{(2)} B_t^2 = 2\, \dot{\nabla}_r B_t \mathbf{1}_{[0,t]}(s)$$

$$= 2\, \mathbf{1}_{[0,t]}(r)\mathbf{1}_{[0,t]}(s).$$

We thus obtain,

$$B_t^2 = t + \frac{1}{2} J_2^s(2\mathbf{1}_{[0,t]} \otimes \mathbf{1}_{[0,t]})$$

$$= t + 2\, J_2(\mathbf{1}_{[0,t]^2}\mathbf{1}_{\mathcal{T}_2})$$

$$= t + 2 \int_0^t \left(\int_0^r dB(s)\right) dB(r)$$

$$= t + 2 \int_0^t B(r) dB(r).$$

We retrieve the well known formula that may be obtained by the Itô formula.

The very same method of identification can be used to prove the next results.

Lemma 3.2 *The vector space spanned by the pure tensors:*

$$\text{span}\{\dot{h}^{\otimes n}, \ \dot{h} \in L^2([0,1] \to \mathbf{R}; \ \ell)\}$$

is dense in $L_s^2([0,1]^n \to \mathbf{R}; \ \ell^{\otimes n})$.

Proof We already know (see Theorem 2.10) that tensor products $\dot{h}_1 \otimes \ldots \otimes \dot{h}_n$ with $\dot{h}_i \in L^2([0,1] \to \mathbf{R}; \ \ell)$ are dense in $L^2([0,1]^n \to \mathbf{R}; \ \ell^{\otimes n})$ and that the symmetrization operation is continuous from $L^2([0,1]^n \to \mathbf{R}; \ \ell)$ into $L_s^2([0,1]^n \to \mathbf{R}; \ \ell^{\otimes n})$. Apply the symmetrization to any approximating sequence to obtain a sequence of linear combinations of pure tensors, which converges to the symmetrization of f_n, which is already f_n. □

Definition 3.4 For A a continuous linear map from $\mathcal{H}$ into itself, we denote by Γ_A the map defined by

$$\Gamma_A : \mathfrak{F}_\mu(\mathcal{H}) \longrightarrow \mathfrak{F}_\mu(\mathcal{H})$$

$$h_1 \otimes \ldots \otimes h_n \longmapsto Ah_1 \otimes \ldots \otimes Ah_n$$

and extended by density to $\mathfrak{F}_\mu(\mathcal{H})$.

Theorem 3.6 (Gradient and Conditional Expectation) *For any* $t \in [0, 1]$, *for any* $F \in L^2(W \to \mathbf{R}; \mu)$,

$$\mathbf{E}[F \mid \mathcal{F}_t] = \mathbf{E}[F] + \sum_{n=1}^{\infty} \frac{1}{n!} \delta^n \left(\Gamma_{\pi_t} \Upsilon_n F \right) \tag{3.13}$$

where we recall that π_t *is the projection map*

$$\pi_t : \mathcal{H} \longrightarrow \mathcal{H}$$
$$h \longmapsto I^1(\dot{h} \, \mathbf{1}_{[0,t]}).$$

Proof The well known identity

$$\mathbf{E}\left[\exp\left(\int_0^1 \dot{h}(s) \mathrm{d}B(s) - \frac{1}{2} \int_0^1 \dot{h}(s)^2 \, \mathrm{d}s \right) \mid \mathcal{F}_t \right]$$
$$= \exp\left(\int_0^t \dot{h}(s) \mathrm{d}B(s) - \frac{1}{2} \int_0^t \dot{h}(s)^2 \mathrm{d}s \right)$$

can be written as

$$\mathbf{E}[\Lambda_h \mid \mathcal{F}_t] = \Lambda_{\pi_t h}.$$

Apply this equality to τh and consider the chaos expansion of both terms. Since the convergence of the series holds in $L^2(W \to \mathbf{R}; \mu)$, we can apply Fubini's theorem straightforwardly.

$$1 + \sum_{n=1}^{\infty} \frac{\tau^n}{n!} \mathbf{E}\left[\delta^n h^{\otimes n} \mid \mathcal{F}_t \right] = 1 + \sum_{n=1}^{\infty} \frac{\tau^n}{n!} \delta^n (\pi_t^{\otimes n} h^{\otimes n}).$$

This means that (3.13) holds for $F \in \mathcal{E}$ and by density, it is true for any $F \in L^2(W \to \mathbf{R}; \mu)$. □

Remark 3.1 (Fundamental Theorem of Calculus Revisited) The fundamental theorem of calculus says that

$$f(t) = f(0) + \int_0^1 f'(rt) \, t \mathrm{d}r.$$

The so-called Clark formula plays the same rôle in the context of stochastic integrals.

Theorem 3.7 (Clark–Ocone Formula) *The map*

$$\partial_W \; : \; \mathcal{E} \longrightarrow L^2(W \to \mathbf{R}; \; \mu)$$

$$F \longmapsto \int_0^1 \mathbf{E}\left[\dot{\nabla}_s F \,|\, \mathcal{F}_s\right] dB(s)$$

can be extended as a continuous map from $L^2(W \to \mathbf{R}; \; \mu)$ into $L^2(W \to \mathbf{R}; \; \mu)$. Moreover,

$$F = \mathbf{E}[F] + \partial_W F. \tag{3.14}$$

For $F \in \mathbb{D}_{1,2}$, this boils down to

$$F = \mathbf{E}[F] + \int_0^1 \mathbf{E}\left[\dot{\nabla}_s F \,|\, \mathcal{F}_s\right] dB(s). \tag{3.15}$$

It is well known that a Brownian martingale can be represented as a stochastic integral with respect to the said Brownian motion but the proof is not constructive and the integrand that has to be considered is defined by a limit procedure from which we cannot devise its value. The Clarke–Ocone formula fills this void and gives the expression of this mysterious process. Actually, the proof of the Clark–Ocone formula proceeds along the same lines as the proof of the martingale representation theorem: establish the validity of the representation for Doléans-Dade exponentials and then pass to the limit. The added value of the Malliavin calculus is that we can express the integrand in an intrinsic way, i.e., as $\mathbf{E}\left[\dot{\nabla}_s F \,|\, \mathcal{F}_s\right]$, which is still well defined even after the limit is taken.

Proof **Step 1** For $F = \Lambda_k$,

$$\partial_W F = \int_0^1 \Lambda_k(s)\dot{k}(s) dB(s) = \int_0^1 \mathbf{E}\left[\dot{\nabla}_s F \,|\, \mathcal{F}_s\right] dB(s)$$

and

$$F = \mathbf{E}[F] + \partial_W F. \tag{3.16}$$

By linearity this remains valid for $F \in \mathcal{E}$.

Step 2 Then,

$$\mathbf{E}\left[\partial_W F^2\right] = \mathbf{E}\left[(F-1)^2\right] = \mathbf{E}\left[F^2\right] - \mathbf{E}[F]^2 \leq \mathbf{E}\left[F^2\right]. \tag{3.17}$$

Step 3 Let $F \in L^2(W \to \mathbf{R}; \mu)$ be the limit of $(F_n, n \geq 1)$ a sequence of elements of $\mathcal{E}$. Equation (3.17) implies that $(\partial_W F_n, n \geq 1)$ is Cauchy in $L^2(W \to \mathbf{R}; \mu)$, hence convergent to a limit, we define to be $\partial_W F$.

Step 4 Then, (3.14) follows from (3.16) by density.

$\square$

As polynomials behave well with derivation, so do the chaos for the Malliavin derivative.

Theorem 3.8 (Gradient of Chaos) *For $\dot{h}_n \in L^2_s([0, 1]^n \to \mathbf{R}, \ell)$, let $\dot{h}(., r)$ be the element of $L^2_s([0, 1]^{n-1} \to \mathbf{R}, \ell)$ defined by*

$$\dot{h}_n(., r) : [0, 1]^{n-1} \longrightarrow \mathbf{R}$$

$$(s_1, \cdots, s_{n-1}) \longmapsto \dot{h}_n(s_1, \cdots, s_{n-1}, r).$$

Then,

$$\dot{\nabla}_r J_n^s(\dot{h}_n) = n \, J_{n-1}^s(\dot{h}_n(., r)). \tag{3.18}$$

Proof Step 1 In view of Lemma 3.2, it is sufficient to prove (3.18) for $\dot{h}_n = \dot{h}^{\otimes n}$. It boils down to prove

$$\dot{\nabla}_r J_n^s(\dot{h}^{\otimes n}) = n \, J_{n-1}^s(\dot{h}^{\otimes n-1})\dot{h}.$$

Let $h \in \mathcal{H}$, we already know that Λ_h belongs to $\mathbb{D}_{1,2}$ and that $\nabla \Lambda_h = \Lambda_h \, h$. Apply this reasoning to τh:

$$\Lambda_h \, h = \sum_{n=0}^{\infty} \frac{\tau^{n+1}}{n!} \, J_n^s(\dot{h}^{\otimes n}) \, h$$

$$= \sum_{n=1}^{\infty} \frac{\tau^n}{(n-1)!} \, J_{n-1}^s(\dot{h}^{\otimes n-1}) \, h$$

$$= \sum_{n=1}^{\infty} \frac{\tau^n}{n!} \, n \, J_{n-1}^s(\dot{h}^{\otimes n-1}) \, h. \tag{3.19}$$

Step 2 We cannot show directly that we can differentiate term by term the chaos expansion of Λ_h, but we can do it in a weak sense: If U belongs to $\mathbb{D}_{1,2}(\mathcal{H})$,

$$\mathbf{E}\left[\left\langle \nabla \left(\sum_{n=0}^{\infty} \frac{\tau^n}{n!} J_n^s(\dot{h}^{\otimes n})\right), U \right\rangle_{\mathcal{H}}\right] = \mathbf{E}\left[\sum_{n=0}^{\infty} \frac{\tau^n}{n!} \langle \nabla J_n^s(\dot{h}^{\otimes n}), U \rangle_{\mathcal{H}}\right]. \tag{3.20}$$

To prove this identity, consider

$$\Lambda_h^{(N)} = \sum_{n=0}^{N} \frac{1}{n!} J_n^s(\dot{h}^{\otimes n}).$$

It holds that

$$\Lambda_h^{(N)} \xrightarrow[L^2(W \to \mathbf{R}; \ell)]{N \to \infty} \Lambda_h.$$

Consequently, $(\nabla \Lambda_h^{(N)}, n \geq 1)$ converges weakly in $\mathbb{D}_{1,2}(\mathcal{H})$ to $\nabla \Lambda_h$: For $U \in \mathbb{D}_{1,2}(\mathcal{H}) \subset \mathrm{Dom}_2 \, \delta$,

$$\mathbf{E}\left[\left\langle \nabla \Lambda_h^{(N)}, U \right\rangle_{\mathcal{H}}\right] = \mathbf{E}\left[\Lambda_h^{(N)} \delta U\right] \xrightarrow{N \to \infty} \mathbf{E}[\Lambda_h \, \delta U] = \mathbf{E}\left[\langle \nabla \Lambda_h, U \rangle_{\mathcal{H}}\right].$$

Furthermore,

$$\mathbf{E}\left[\left\langle \nabla \Lambda_{\tau h}^{(N)}, U \right\rangle_{\mathcal{H}}\right] = \sum_{n=1}^{N} \frac{\tau^n}{n!} \mathbf{E}\left[\langle \nabla J_n^s(\dot{h}^{\otimes n}), U \rangle_{\mathcal{H}}\right],$$

so (3.20) is satisfied.

Step 3. In view of (3.19), we also have

$$\mathbf{E}\left[\left\langle \nabla \left(\sum_{n=0}^{\infty} \frac{\tau^n}{n!} J_n^s(\dot{h}^{\otimes n}) \right), U \right\rangle_{\mathcal{H}}\right] = \sum_{n=1}^{\infty} \frac{\tau^n}{n!} \, n \, \mathbf{E}\left[J_{n-1}^s(\dot{h}^{\otimes n-1}) \, \langle h, U \rangle_{\mathcal{H}}\right].$$

Identify the coefficient of τ^n: For any $U \in \mathbb{D}_{1,2}(\mathcal{H})$

$$\mathbf{E}\left[\langle \nabla J_n^s(\dot{h}^{\otimes n}), U \rangle_{\mathcal{H}}\right] = n \, \mathbf{E}\left[\left\langle J_{n-1}^s(\dot{h}^{\otimes n-1}) \, h, U \right\rangle_{\mathcal{H}}\right].$$

Since $\mathbb{D}_{1,2}(\mathcal{H})$ contains the $\mathcal{H}$-valued cylindrical functions that are dense in $L^2(W \to \mathcal{H}; \mu)$, we have

$$\nabla J_n^s(\dot{h}^{\otimes n}) = n \, J_{n-1}^s(\dot{h}^{\otimes n-1}) \, h, \quad \mu - \text{a.s.}$$

and the result follows. $\square$

Corollary 3.1 *A random variable $F \in L^2(W \to \mathbf{R}; \mu)$ belongs to $\mathbb{D}_{2,1}$ if and only if*

$$\sum_{n=1}^{\infty} \frac{1}{(n-1)!} \|\Upsilon_n F\|_{L^2([0,1]^n)}^2 < \infty, \tag{3.21}$$

and ∇F is given by

$$\dot{\nabla}_r F = \sum_{n=1}^{\infty} \frac{1}{(n-1)!} J_{n-1}^s (\Upsilon_n F(., r)). \tag{3.22}$$

Proof Step 1 For any $N > 0$, let

$$F_N = \sum_{n=0}^{N} \frac{1}{n!} J_n^s (\Upsilon_n F).$$

According to the previous theorem, we have

$$\dot{\nabla}_r \left(\sum_{n=0}^{N} \frac{1}{n!} J_n^s (\Upsilon_n F) \right) = \sum_{n=1}^{N} \frac{1}{(n-1)!} J_{n-1}^s (\Upsilon_n F(., r)).$$

Step 2 In view of (3.6), we get

$$\mathbf{E} \left[\int_0^1 \dot{\nabla}_r \left(\sum_{n=0}^{N} \frac{1}{n!} J_n^s (\Upsilon_n F) \right)^2 dr \right]$$

$$= \sum_{n=1}^{N} \frac{(n-1)!}{(n-1)!^2} \int_{[0,1]^n} (\Upsilon_n F)(r_1, \cdots, r_{n-1}, r)^2 dr_1 \ldots dr$$

$$= \sum_{n=1}^{N} \frac{1}{(n-1)!} \| \Upsilon_n F(., r) \|_{L^2([0,1]^n \to \mathbf{R}; \ell)}^2. \tag{3.23}$$

Step 3 If $F \in \mathbb{D}_{2,1}$, then

$$\dot{\nabla}_r F_N \xrightarrow[L^2(\mathsf{W} \times [0,1] \to \mathbf{R}; \mu \otimes \ell)]{N \to \infty} \dot{\nabla}_r F$$

hence the right-hand side of (3.23) converges and (3.21) is satisfied.

Step 4 Conversely, assume that the right-hand side of (3.23) converges. This means that $\sup_N \| F_N \|_{\mathbb{D}_{2,1}} < \infty$ and according to Lemma 2.3, F belongs to $\mathbb{D}_{2,1}$ and its gradient is given by (3.22). $\qquad\square$

Definition 3.5 For $\dot{f} \in L^2([0, 1]^n \to \mathbf{R}; \ell)$ and $\dot{g} \in L^2([0, 1]^m \to \mathbf{R}; \ell)$, for $i \leq n \wedge m$, the i-th contraction of $\dot{f}$ and $\dot{g}$ is defined by

$$(\dot{f} \otimes_i \dot{g})(t_1, \cdots, t_{n-i}, s_1, \cdots, s_{m-i})$$

$$= \int_{[0,1]^i} \dot{f}(t_1, \cdots, t_{n-i}, u_1, \cdots, u_i) \, \dot{g}(s_1, \cdots, s_{m-i}, u_1, \cdots, u_i) du_1 \ldots du_i.$$

It is an element of $L^2([0, 1]^{n+m-2i} \to \mathbf{R}; \ell)$. Its symmetrization is denoted by $\dot{f} \overset{s}{\otimes}_i g$. By convention, $\dot{f} \otimes_0 \dot{g} = \dot{f} \otimes \dot{g}$ and if $n = m$, $\dot{f} \otimes_n \dot{g} = \langle f, g \rangle_{\mathcal{H}^{\otimes(n)}}$.

Theorem 3.9 (Multiplication of Iterated Integrals) *For* $\dot{f} \in L^2([0, 1]^n \to \mathbf{R}; \ell)$ *and* $\dot{g} \in L^2([0, 1]^m \to \mathbf{R}; \ell)$,

$$J_n^s(\dot{f}) J_m^s(\dot{g}) = \sum_{i=0}^{n \wedge m} \frac{n!m!}{i!(n-i)!(m-i)!} \, J_{n+m-2i}(\dot{f} \overset{s}{\otimes}_i \dot{g}). \tag{3.24}$$

Proof We give the proof for $n = 1$; the general case follows the same principle with much involved notations and computations. Without loss of generality, we can assume $\dot{g}$ symmetric.

For $\psi \in \mathcal{E}$,

$$\mathbf{E}\left[J_m^s(\dot{g}) J_1^s(\dot{f}) \psi\right] = \mathbf{E}\left[\delta^m(g) \, \delta f \psi\right] = \mathbf{E}\left[\left\langle \nabla^{(m)}(\psi \, \delta f), \, g \right\rangle_{\mathcal{H}^{\otimes m}}\right].$$

Recall that $\nabla \delta f = f$ and that $\nabla^k \delta f = 0$ if $k \geq 2$. The Leibniz formula then implies that

$$\nabla^{(m)}(\psi \, \delta f) = \delta f \, \nabla^{(m)} \psi + m \, \nabla^{(m-1)} \psi \otimes f.$$

On the one hand,

$$\mathbf{E}\left[\delta f \, \left\langle \nabla^{(m)} \psi, \, g \right\rangle_{\mathcal{H}^{\otimes m}}\right] = \mathbf{E}\left[\left\langle \nabla^{(m+1)} \psi, \, g \otimes f \right\rangle_{\mathcal{H}^{\otimes(m+1)}}\right]$$

$$= \mathbf{E}\left[\psi \, \delta^{m+1}(g \otimes f)\right].$$

On the other hand, a simple application of Fubini's Theorem yields

$$\mathbf{E}\left[\left\langle \nabla^{(m-1)} \psi \otimes f, \, g \right\rangle_{\mathcal{H}^{\otimes m}}\right]$$

$$= \mathbf{E}\left[\int_{[0,1]^m} \dot{\nabla}_{s_1, \cdots, s_{m-1}}^{(m-1)} \psi \, \dot{f}(s_m) \, \dot{g}(s_1, \cdots, s_m) ds_1 \ldots ds_m\right]$$

$$= \mathbf{E}\left[\int_{[0,1]^{m-1}} \dot{\nabla}_{s_1, \cdots, s_{m-1}}^{(m-1)} \psi \, (\dot{f} \otimes_1 \dot{g})(s_1, \cdots, s_{m-1}) ds_1 \ldots ds_{m-1}\right].$$

Since g is symmetric with respect to its $(m-1)$ first variables, the function $\dot{f} \otimes_1 \dot{g}$ is symmetric hence equals to $\dot{f} \overset{s}{\otimes}_1 \dot{g}$.

Finally, we get

$$\mathbf{E}\left[\left\langle \dot{\nabla}^{(m-1)}\psi \otimes f,\ g\right\rangle_{\mathcal{H}^{\otimes m}}\right] = \mathbf{E}\left[\psi\ \delta^{m-1}(\dot{f} \overset{s}{\otimes}_1 \dot{g})\right].$$

The result follows by the density of $\mathcal{E}$ in $L^2(\mathrm{W} \to \mathbf{R};\ \mu)$.   □

Corollary 3.2 (Divergence on Chaos) *If $\dot{U}$ admits the representation*

$$\dot{U}(t) = J_n^s\left(\dot{h}_n(.,t)\right)$$

where $\dot{h}_n$ belongs to $L^2([0,1]^{n+1} \to \mathbf{R};\ \ell)$ and is symmetric with respect to its n first variables. Then, we have

$$\delta U = J_{n+1}^s(\tilde{h}_n)$$

where

$$\tilde{h}_n(t_1,\cdots,t_n,t_{n+1}) = \frac{1}{n+1}\left[\dot{h}_n(t_1,\cdots,t_n,t_{n+1})\right.$$

$$\left. + \sum_{i=1}^{n}\dot{h}_n(t_1,\cdots,t_{i-1},t_{n+1},t_{i+1},\cdots,t_i)\right]. \quad (3.25)$$

Proof As before, we reduce the problem to $\dot{h}_n(.,t) = \dot{h}^{\otimes n}\dot{g}(t)$. Then,

$$J_n^s(\dot{h}^{\otimes n}\dot{g}(t)) = J_n^s(\dot{h}^{\otimes n})\dot{g}(t).$$

Equation (2.27), (3.24), and (3.18) imply

$$\delta(J_n^s(\dot{h}^{\otimes n})\,g) = J_n^s(\dot{h}^{\otimes n})J_1(\dot{g}) - \left\langle \nabla J_n^s(\dot{h}^{\otimes n}),\ g\right\rangle_{\mathcal{H}}$$

$$= J_{n+1}^s(\dot{h}^{\otimes n} \overset{s}{\otimes} \dot{g}) + nJ_{n-1}^s(\dot{h}^{\otimes n} \overset{s}{\otimes}_1 \dot{g}) - nJ_{n-1}^s(\dot{h}^{\otimes n-1})\int_0^1 h(s)\dot{g}(s)\mathrm{d}s.$$

$$(3.26)$$

By its very definition,

$$\left(\dot{h}^{\otimes n} \overset{s}{\otimes}_1 \dot{g}\right)(t_1,\cdots,t_{n-1}) = \prod_{j=1}^{n-1}\dot{h}(t_j)\int_0^1 \dot{h}(s)\dot{g}(s)\mathrm{d}s,$$

hence the last two terms of (3.26) do cancel each other. Since $\dot{h}^{\otimes n-1}$ is already symmetric, the symmetrization of $\dot{h}^{\otimes n-1} \otimes \dot{g}$ reduces to

$$\left(\dot{h}^{\otimes n-1} \overset{s}{\otimes} \dot{g}\right)(t_1, \cdots, t_{n+1})$$

$$= \frac{1}{n+1}\left[\prod_{j=1}^{n} \dot{h}(t_j)\dot{g}(t_{n+1}) + \sum_{i=1}^{n}\prod_{\substack{j=1 \\ j \neq i}}^{n} \dot{h}(t_j)\dot{g}(t_i)\dot{h}(t_{n+1})\right],$$

which corresponds to (3.26) for general $\dot{h}_n$. □

3.2 Ornstein–Uhlenbeck Operator

In $\mathbf{R}^n$, the adjoint of the usual gradient is the divergence operator and the composition of divergence and gradient is the ordinary Laplacian. Since we have at our disposal, a notion of gradient and the corresponding divergence, we can consider the associated Laplacian, sometimes called Gross Laplacian, defined as

$$L = \delta\nabla.$$

A simple calculation shows the following that justifies the physicists' denomination of L as the number operator.

Chaos can be defined as iterated integrals. In a more functional manner, they can be seen as the eigenfunctions of L.

Theorem 3.10 (Number Operator) *Let $F \in L^2(\mathbf{W} \to \mathbf{R}; \mu)$ of chaos decomposition*

$$F = \mathbf{E}[F] + \sum_{n=1}^{\infty} J_n^s(\dot{h}_n).$$

We say that F belongs to Dom L *whenever*

$$\sum_{n=1}^{\infty} n^2 \|J_n^s(\dot{h}_n)\|^2_{L^2(\mathbf{W} \to \mathbf{R}; \mu)} < \infty.$$

Then, for such an F, we have

$$LF = \sum_{n=1}^{\infty} n\, J_n^s(\dot{h}_n).$$

The map L is invertible from $L_0^2 = \left\{ F \in L^2(W \to \mathbf{R}; \mu),\ \mathbf{E}[F] = 0 \right\}$ *into itself:*

$$L^{-1}F = \sum_{n=1}^{\infty} \frac{1}{n} J_n^s(\dot{h}_n).$$

From there, it is customary to define the so-called Ornstein–Uhlenbeck operator from its action on chaos.

Definition 3.6 (Ornstein–Uhlenbeck Operator) Let $F \in L^2(W \to \mathbf{R}; \mu)$ of chaos decomposition

$$F = \mathbf{E}[F] + \sum_{n=1}^{\infty} J_n^s(\dot{h}_n).$$

For any $t > 0$,

$$P_t F = \mathbf{E}[F] + \sum_{n=1}^{\infty} e^{-nt} J_n^s(\dot{h}_n).$$

Formally, we can write $P_t = e^{-tL}$.

From these definitions, the following properties are straightforward.

Theorem 3.11 *For any* $F \in L^2(W \to \mathbf{R}; \mu)$, *for any* $s, t \geq 0$,

$$P_{t+s} F = P_s(P_t F).$$

For any $F \in \mathbb{D}_{2,1}$,

$$\nabla P_t F = e^{-t} P_t \nabla F. \tag{3.27}$$

The Ornstein–Uhlenbeck can be alternatively defined by the so-called Mehler formula:

Theorem 3.12 *For any* $F \in L^2(W \to \mathbf{R}; \mu)$

$$P_t F(\omega) = \int_W F\left(e^{-t}\omega + \sqrt{1 - e^{-2t}}\, y\right) d\mu(y). \tag{3.28}$$

The Mehler formula presents the Ornstein–Uhlenbeck as a sort of convolution operator. As such, it will benefit from strong regularizing properties that are often useful for approximation procedures.

Part of the theorem consists in proving that the integral is well defined. Actually, the law of $\omega + B$ is singular with respect to the law of B whenever ω does not belong to $\mathcal{H}$ hence as such, the right-hand side of (3.28) is not properly defined for a measurable only F. We are going to prove that it is unambiguously defined when F belongs to $\mathcal{E}$ and then define the integral by density thanks to an invariance property of the Wiener measure.

In what follows, let $\beta_t = \sqrt{1 - e^{-2t}}$.

Lemma 3.3 *For any $t > 0$, consider the transformation*

$$R_t : W \times W \longrightarrow W \times W$$

$$(\omega, \eta) \longmapsto \left(e^{-t}\omega + \beta_t \eta, \; -\beta_t \omega + e^{-t}\eta\right).$$

Then the image of $\mu \otimes \mu$ by R_t is still $\mu \otimes \mu$.

Proof Let h_1 and h_2 belong to W^*. Then,

$$\int_{W \times W} \exp\left(i \langle (h_1, h_2), R_t(\omega, \eta)\rangle_{W^* \times W^*, W \times W}\right) d\mu(\omega)\, d\mu(\eta)$$

$$= \int_W \exp\left(i \langle e^{-t}h_1 - \beta_t h_2, \omega\rangle_{W^*, W}\right) d\mu(\omega)$$

$$\times \int_W \exp\left(i \langle e^{-t}h_2 + \beta_t h_1, \eta\rangle_{W^*, W}\right) d\mu(\eta)$$

$$= \exp\left(-\frac{1}{2}\left(\|e^{-t}h_1 - \beta_t h_2\|_{\mathcal{H}}^2 + \|e^{-t}h_2 + \beta_t h_1\|_{\mathcal{H}}^2\right)\right)$$

$$= \exp\left(-\frac{1}{2}\|h_1\|_{\mathcal{H}}^2\right) \exp\left(-\frac{1}{2}\|h_2\|_{\mathcal{H}}^2\right).$$

In view of the characterization of the Wiener measure, this completes the proof. □

Proof of Theorem 3.12 **Step 1** For $h \in \mathcal{H}$,

$$\delta h\left(e^{-t}\omega + \beta_t \eta\right) = \delta(e^{-t}h)(\omega) + \delta(\beta_t h)(\eta)$$

and

$$\|h\|_{\mathcal{H}}^2 = \|e^{-t}h\|_{\mathcal{H}}^2 + \|\beta_t h\|_{\mathcal{H}}^2.$$

Hence,

$$\Lambda_h\left(e^{-t}\omega + \beta_t \eta\right) = \Lambda_{e^{-t}h}(\omega)\Lambda_{\beta_t h}(\eta).$$

It follows that

$$\int_W \Lambda_h\left(e^{-t}\omega + \beta_t \eta\right) d\mu(\eta) = \Lambda_{e^{-t}h}(\omega) \int_W \Lambda_{\beta_t h}(\eta) d\mu(\eta) = \Lambda_{e^{-t}h}(\omega).$$

Now then, the chaos decomposition of $\Lambda_{e^{-t}h}(\omega)$ is given by

$$\Lambda_{e^{-t}h}(\omega) = 1 + \sum_{n=1}^{\infty} \frac{1}{n!} J_n^s\big((e^{-t}h)^{\otimes n}\big) = 1 + \sum_{n=1}^{\infty} \frac{e^{-nt}}{n!} J_n^s\big(\dot{h}^{\otimes n}\big).$$

We have thus proved that

$$\int_W F\left(e^{-t}\omega + \sqrt{1 - e^{-2t}}\, y\right) d\mu(y)$$

is well defined for $F \in \mathcal{E}$ and that (3.28) holds for such functionals F.

Step 2 For $F \in \mathcal{E}$, We have

$$\int_W \left(\int_W F\left(e^{-t}\omega + \sqrt{1 - e^{-2t}}\, y\right) d\mu(y)\right)^2 d\mu(\omega)$$

$$\leq \int_{W^2} F(e^{-t}\omega + \sqrt{1 - e^{-2t}}\, \eta)^2 d\mu(\omega) d\mu(\eta)$$
$$= \int_{W^2} \bar{F}\big(R_t(\omega, \eta)\big)^2 d\mu(\omega) d\mu(\eta),$$

where

$$\bar{F} : W \times W \longrightarrow \mathbf{R}$$

$$(\omega, \eta) \longmapsto F(\omega).$$

According to Lemma 3.3,

$$\int_{W^2} \bar{F}\big(R_t(\omega, \eta)\big)^2 d\mu(\omega)\, d\mu(\eta) = \int_{W^2} \bar{F}(\omega, \eta)^2 d\mu(\omega) d\mu(\eta)$$

$$= \|F\|^2_{L^2(W \to \mathbf{R};\, \mu)},$$

or equivalently

$$\left\| \int_W F\left(e^{-t}\omega + \sqrt{1 - e^{-2t}}\, y\right) d\mu(y) \right\|_{L^2(W \to \mathbf{R};\, \mu)} \leq \|F\|_{L^2(W \to \mathbf{R};\, \mu)}. \tag{3.29}$$

Thus, by a density argument, we can extend the integral to the whole of $L^2(W \to \mathbf{R};\, \mu)$.

Step 3 We know that for $F \in L^2(W \to \mathbf{R};\, \mu)$,

$$\sum_{n=1}^{\infty} \frac{1}{n!} \mathbf{E}\big[J_n^s(\dot{h}_n)^2 \big] < \infty.$$

If each kernel is multiplied by a constant smaller than 1, the convergence also holds, hence for any $t \geq 0$,

$$\|P_t F\|_{L^2(W \to \mathbf{R};\, \mu)} \leq \|F\|_{L^2(W \to \mathbf{R};\, \mu)}.$$

We can then conclude by a density argument.

$\square$

The Ornstein–Uhlenbeck semi-group can also be seen as the semi-group associated to a W-valued Markov process whose generator would be formally L, see (3.36) for details. As such, it is interesting to note that the stationary distribution of this Markov process is the Wiener measure.

Theorem 3.13 *The semi-group is ergodic and admits μ as stationary measure. As a consequence,*

$$\int_W F d\mu - F = -\int_0^\infty L P_t F d\mu \qquad (3.30)$$

and for F centered,

$$L^{-1} F = \int_0^\infty P_t F dt. \qquad (3.31)$$

Proof From the Mehler formula, we see by dominated convergence that

$$P_t F(\omega) \xrightarrow[\text{w.p.1}]{t \to \infty} \int_W F d\mu.$$

In view of Lemma 3.3,

$$\int_W P_t F(\omega) d\mu(\omega) = \int_{W^2} \bar{F}\big(R_t(\omega,\, y)\big) d\mu(\omega) d\mu(y)$$

$$= \int_{W^2} \bar{F}(\omega,\, y) d\mu(\omega) d\mu(y) = \int_W F(\omega) d\mu(\omega).$$

This proves the stationarity of μ. Now, it comes from the chaos decomposition that

$$\frac{d}{dt} P_t F = -L P_t F,$$

hence

$$P_t F(\omega) - P_0 F(\omega) = -\int_0^t L P_t F(\omega) dt.$$

Let t go to infinity to obtain (3.30). Equation (3.31) is a direct consequence of the chaos decomposition. □

The Mehler formula shows that $P_t F$ is a convolution operator and as such has some strong regularization properties.

Definition 3.7 (Generalized Hermite Polynomials) The generalized Hermite polynomials are defined by their generating function:

$$\exp(\alpha x - \frac{\alpha^2}{2}t) = \sum_{n=0}^{\infty} \frac{\alpha^n}{n!} \mathfrak{H}_n(x,t).$$

We have

$$\mathfrak{H}_0(x,t) = 1, \quad \mathfrak{H}_1(x,t) = x, \quad \mathfrak{H}_2(x,t) = x^2 - t.$$

The usual Hermite polynomials correspond to $\mathfrak{H}_n(x,1)$.

Theorem 3.14 (Regularization) *For $F \in L^p(\mathbf{W} \to \mathbf{R}; \mu)$, for any $t > 0$, $P_t F$ belongs to $\cap_{k \geq 1} \mathbb{D}_{k,p}$. Moreover,*

$$\left\langle \nabla^{(k)} P_t F, h^{\otimes k} \right\rangle_{\mathcal{H}^{\otimes k}} = \left(\frac{e^{-t}}{\beta_t} \right)^k \int_{\mathbf{W}} F(e^{-t}\omega + \beta_t y)\, \mathfrak{H}_k(\delta h(y), \|h\|_{\mathcal{H}}^2) d\mu(y). \tag{3.32}$$

Proof **Step 1** For $k = 1$, for $F \in \mathcal{S}$,

$$\langle \nabla P_t F, h \rangle_{\mathcal{H}} = \frac{d}{d\varepsilon} P_t F(\omega + \varepsilon h) \Big|_{\varepsilon = 0}$$

The trick is then to consider that the translation by h operates not on ω but on y:

$$P_t F(\omega + \varepsilon h) = \int_{\mathbf{W}} F(e^{-t}(\omega + \varepsilon h) + \beta_t y) d\mu(y)$$

$$= \int_{\mathbf{W}} F(e^{-t}\omega + \beta_t(y + \frac{\varepsilon e^{-t}}{\beta_t} h)) d\mu(y).$$

According to the Cameron–Martin (Theorem 1.8),

$$P_t F(\omega + \varepsilon h) = \int_{\mathbf{W}} F(e^{-t}\omega + \beta_t y) \exp\left(\varepsilon \frac{e^{-t}}{\beta_t} \delta h - \frac{\varepsilon^2 e^{-2t}}{\beta_t^2} \|h\|_{\mathcal{H}}^2 \right) d\mu(y).$$

Since,

$$\frac{d}{d\varepsilon} \left(\varepsilon \frac{e^{-t}}{\beta_t} \delta h - \frac{\varepsilon^2 e^{-2t}}{\beta_t^2} \|h\|_{\mathcal{H}}^2 \right) \Big|_{\varepsilon = 0} = \frac{e^{-t}}{\beta_t} \delta h,$$

the result follows by dominated convergence.

Step 2 For $k = 2$, we proceed along the same lines

$$\langle \nabla P_t F(\omega + \varepsilon h), h \rangle_{\mathcal{H}} = \frac{e^{-t}}{\beta_t} \int_W F\left(e^{-t}\omega + \beta_t\left(y + \frac{\varepsilon e^{-t}}{\beta_t}h\right)\right) \delta h(y) d\mu(y)$$

$$= \frac{e^{-t}}{\beta_t} \int_W F\left(e^{-t}\omega + \beta_t y\right) \delta h\left(y - \frac{\varepsilon e^{-t}}{\beta_t}h\right) \Lambda_{\varepsilon \frac{e^{-t}}{\beta_t}h}(y) d\mu(y)$$

$$= \frac{e^{-t}}{\beta_t} \int_W F\left(e^{-t}\omega + \beta_t y\right)\left(\delta h(y) - \frac{\varepsilon e^{-t}}{\beta_t}\|h\|_{\mathcal{H}}^2\right) \Lambda_{\varepsilon \frac{e^{-t}}{\beta_t}h}(y) d\mu(y)$$

Hence,

$$\left\langle \nabla^{(2)} P_t F(\omega), h^{\otimes 2} \right\rangle_{\mathcal{H}} = \frac{d}{d\varepsilon} \langle \nabla P_t F(\omega + \varepsilon h), h \rangle_{\mathcal{H}} \Big|_{\varepsilon = 0}$$

$$= \left(\frac{e^{-t}}{\beta_t}\right)^2 \int_W F\left(e^{-t}\omega + \beta_t y\right)\left(\delta h(y)^2 - \|h\|_{\mathcal{H}}^2\right) d\mu(y).$$

The formula for general k follows by recursion.

Step 3 For $F \in L^p(W \to \mathbf{R}; \mu)$, let $(F_n, n \geq 1)$ be a sequence of cylindrical functions converging in $L^p(W \to \mathbf{R}; \mu)$ to F. For any $h \in \mathcal{H}$, $\delta h(y)$ is a Gaussian random variable hence $\mathfrak{H}_k(\delta h(y), \|h\|_{\mathcal{H}}^2)$ belongs to $L^q(W \to \mathbf{R}; \mu)$ and we have

$$\left| \int_W F_n(e^{-t}\omega + \beta_t y) \, \mathfrak{H}_k(\delta h(y), \|h\|_{\mathcal{H}}^2) d\mu(y) \right| \leq c_p \|F_n\|_{L^p(W \to \mathbf{R}; \mu)} \|h\|_{\mathcal{H}}^{2k}.$$

Hence,

$$\sup_n \|\nabla^{(k)} F_n\|_{L^p(W \to \mathcal{H}^{\otimes k}; \mu)} < \infty$$

and we conclude by Lemma 2.3.

$\square$

One of the most striking applications of the properties of the Ornstein–Uhlenbeck operator is the Meyer inequalities that merely state that the norm derived from $L^{1/2}$ coincides with the Sobolev norms on $\mathbb{D}_{p,k}$.

Theorem 3.15 (Meyer Inequalities) *For any $p > 1$ and any $k \geq 1$, there exist $c_{p,k}$ and $C_{p,k}$ such that for any $F \in \mathbb{D}_{p,k}$,*

$$c_{p,k} \|F\|_{p,k} \leq \mathbf{E}\left[|(I+L)^{k/2}F|^p\right] \leq C_{p,k} \|F\|_{p,k}.$$

See Equation (2.7) for the definition of the norm on $\mathbb{D}_{p,k}$.

Fig. 3.1 Simulation of a
sample-path of S_N

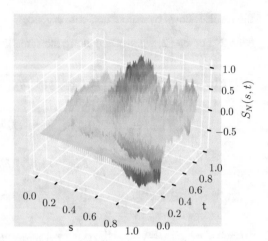

3.3 Problems

3.1 Consider the Brownian sheet W that is the centered Gaussian process indexed
by $[0, 1]^2$ with covariance kernel

$$\mathbf{E}\left[W(t_1, t_2)W(s_1, s_2)\right] = s_1 \wedge t_1 \ s_2 \wedge t_2 := R(s, t). \tag{3.33}$$

Let $(X_{ij}, 1 \leq i, j \leq N)$ be a family of N^2 independent and identically distributed
random variables with mean 0 and variance 1. Define

$$S_N(s, t) = \frac{1}{N} \sum_{i=1}^{[Ns]} \sum_{j=1}^{[Nt]} X_{ij}.$$

See Fig. 3.1 for a simulation of a sample-path of S_N.

1. Show that $\left(S_N(s_1, s_2), S_N(t_1, t_2)\right)$ converges to a Gaussian random vector of
   covariance matrix

$$\Gamma = \begin{pmatrix} R(s, s) & R(s, t) \\ R(s, t) & R(t, t) \end{pmatrix}$$

2. Show that for fixed t, the process $s \longmapsto W(s, \beta_t^2)$ has the same distribution as
   the process $s \longmapsto \beta_t B(s)$ where B is the standard Brownian motion.
3. Derive that for $F \in L^2(\mathrm{W} \to \mathbf{R}; \ \mu)$ and $\omega \in \mathrm{W}$, we have

$$P_t F(\omega) = \mathbf{E}\left[F\left(e^{-t}\omega + W(., \beta_t^2)\right)\right].$$

3.2 From [5], we derive an alternative expression of the second order derivative of
$P_t F$. Assume that F belongs to S and h, k are two elements of $\mathcal{H}$.

1. Use the semi-group property to derive

$$\langle \nabla P_t F(\omega), h \rangle_{\mathcal{H}}$$
$$= \frac{e^{-t/2}}{\beta_{t/2}} \int_{\mathrm{W}} \left(\int_{\mathrm{W}} F\left(e^{-t}\omega + e^{-t/2}\beta_{t/2}y + \beta_{t/2}z\right) d\mu(z) \right) \delta h(y) d\mu(y).$$

2. Show that

$$\left\langle \nabla^{(2)} P_t F(\omega), h \otimes k \right\rangle_{\mathcal{H}}$$
$$= \frac{e^{-3t/2}}{\beta_{t/2}^2} \int_{\mathrm{W}} \int_{\mathrm{W}} F\left(e^{-t}\omega + e^{-t/2}\beta_{t/2}y + \beta_{t/2}z\right) \delta h(y)\delta k(z) d\mu(y) d\mu(z).$$

$$\tag{3.34}$$

3. Show that (3.34) holds for $F \in L^1(\mathrm{W} \to \mathbf{R}; \mu)$.
   Assume that $F \in L^1(\mathrm{W} \to \mathbf{R}; \mu)$ and that F is in $\mathrm{Lip}_1(W)$: For any $\omega' \in \mathrm{W}$,

$$|F(\omega + \omega') - F(\omega)| \le \|\omega'\|_{\mathrm{W}}.$$

4. Use Cauchy–Schwarz inequality to show that

$$\left| \left\langle \nabla^{(2)} P_t F(\omega + \alpha \mathfrak{e}(h)), h \otimes h \right\rangle_{\mathcal{H}} \right| \le \frac{\alpha \, e^{-5t/2}}{\beta_{t/2}^2} \|h\|_{\mathcal{H}}^2 \|\mathfrak{e}(h)\|_{\mathrm{W}}.$$

$$\tag{3.35}$$

3.3 For $f : \mathbf{R}^n \to \mathbf{R}$ twice differentiable, we set

$$L_n f(x) = \langle x, D_n f(x) \rangle_{\mathbf{R}^n} - \Delta_n f(x)$$
$$= \sum_{j=1}^{n} x_j \, \partial_j f(x) - \sum_{j=1}^{n} \partial_{jj}^2 f(x).$$

1. For $F(\omega) = f(\delta h_1(\omega), \cdots, \delta h_n(\omega))$ where $f \in C^2(\mathbf{R}^n, \mathbf{R})$,

$$LF(\omega) = (L_n f)(\delta h_1(\omega), \cdots, \delta h_n(\omega)).$$

$$\tag{3.36}$$

3.4 Notes and Comments

Chaos are interesting because the action of the gradient and of the divergence can be readily seen on each chaos. They also give a convenient definition of the Ornstein–Uhlenbeck semi-group as a diagonal operator. Actually, it can be said that the chaos decomposition plays the rôle of the series expansion for ordinary functions, with the same advantages and limitations.

There are some other probability measures for which the chaos decomposition is known to hold. Beyond the Wiener measure [2, 6], we may consider the distribution of the Poisson process possibly with marks [4], the Rademacher measure that is the distribution of a sequence of independent Bernoulli random variables [4], the distribution of Lévy processes [3], and the distribution of some finite Markov chains [1]. There was a tremendous activity on this subject during the nineties, but to the best of my knowledge, it did not go much farther than these examples.

This version of the proof of the multiplication formula for iterated integrals can be found in [7].

References

1. P. Biane, Chaotic representation for finite Markov chains. Stoch. Stoch. Rep. **30**, 61–68 (1989)
2. D. Nualart, *The Malliavin Calculus and Related Topics* (Springer–Verlag, Berlin, 1995)
3. D. Nualart, W. Schoutens, Chaotic and predictable representations for Lévy processes. Stoch. Process. Appl. **90**(1), 109–122 (2000)
4. N. Privault, *Stochastic Analysis in Discrete and Continuous Settings with Normal Martingales*. Lecture Notes in Mathematics, vol. 1982. (Springer, Berlin, 2009)
5. H.-H. Shih, On Stein's method for infinite-dimensional Gaussian approximation in abstract Wiener spaces. J. Funct. Anal. **261**(5), 1236–1283 (2011)
6. A.S. Üstünel, *An Introduction to Analysis on Wiener Space*. Lectures Notes in Mathematics, vol. 1610 (Springer, Berlin, 1995)
7. A.S. Üstünel, A sophisticated proof of the multiplication formula for multiple wiener integrals (2014). arXiv:1411.4877

Chapter 4
Fractional Brownian Motion

Abstract In the nineties, statistical evidence, notably in finance and telecommunications, showed that Markov processes were too far away from the observations to be considered as viable models. In particular, there were strong suspicions that the data exhibit long range dependence. It is in this context that the fractional Brownian motion, introduced by B. Mandelbrot in the late sixties and almost forgotten since, enjoyed a new rise of interest. It is a Gaussian process with long range dependence. Consequently, it cannot be a semi-martingale, and we cannot apply the theory of Itô calculus. As we have seen earlier, for the Brownian motion, the Malliavin divergence generalizes the Itô integral and can be constructed for the fBm, so it is tempting to view it as an ersatz of a stochastic integral. Actually, the situation is not that simple and depends on what we call a stochastic integral.

4.1 Definition and Sample-Paths Properties

Definition 4.1 For any H in $(0, 1)$, the fractional Brownian motion of index (Hurst parameter) H, $\{B_H(t);\ t \in [0, 1]\}$ is the centered Gaussian process whose covariance kernel is given by

$$R_H(s, t) = \mathbf{E}\left[B_H(s)B_H(t)\right] = \frac{V_H}{2}\left(s^{2H} + t^{2H} - |t - s|^{2H}\right)$$

where

$$V_H = \frac{\Gamma(2 - 2H)\cos(\pi H)}{\pi H(1 - 2H)}.$$

Note that for $H = 1/2$, we obtain

$$R_{1/2}(t, s) = \frac{1}{2}(t + s - |t - s|)$$

L. Decreusefond, *Selected Topics in Malliavin Calculus*,
Bocconi & Springer Series 10, https://doi.org/10.1007/978-3-031-01311-9_4

which is nothing but the sophisticated way to write $R_{1/2}(t, s) = \min(t, s)$. Hence, $B_{1/2}$ is the ordinary Brownian motion.

Theorem 4.1 *Let $H \in (0, 1)$, and the sample-paths of W^H are Hölder continuous of any order less than H (and no more) and belong to $W_{\alpha,p}$ for any $p \geq 1$ and any $\alpha \in (0, H)$.*

We denote by μ_H the measure on $W_{\alpha,p}$ that corresponds to the distribution of B_H.

Proof Step 1 A simple calculation shows that, for any $p \geq 0$, we have

$$\mathbf{E}\left[|B_H(t) - B_H(s)|^p\right] = C_p|t - s|^{Hp}.$$

Since B_H is Gaussian, its p-th moment can be expressed as a power of the variance; hence, we have

$$\mathbf{E}\left[\iint_{[0,1]^2} \frac{|B_H(t) - B_H(s)|^p}{|t - s|^{1+\alpha p}} \, dt \, ds\right] = C_\alpha \iint_{[0,1]^2} |t - s|^{-1+p(H-\alpha)} \, dt \, ds.$$

This integral is finite as soon as $\alpha < H$; hence, for any $\alpha < H$, any $p \geq 1$, B_H belongs to $W_{\alpha,p}$ with probability 1. Choose p arbitrary large and conclude that the sample-paths are Hölder continuous of any order less than H, in view of the Sobolev embeddings (see Theorem 1.4).

Step 2 As a consequence of the results in [1], we have

$$\mu_H\left(\limsup_{u \to 0^+} \frac{B_H(u)}{u^H \sqrt{\log \log u^{-1}}} = \sqrt{V_H}\right) = 1.$$

Hence, it is impossible for B_H to have sample-paths Hölder continuous of an order greater than H.   □

The difference of regularity is evident on simulations of sample-paths, see Fig. 4.1.

Lemma 4.1 *The process $(a^{-H}B_H(at), t \geq 0)$ has the same distribution as B_H.*

Proof Consider the centered Gaussian process

$$Z(t) = a^{-H}B_H(at).$$

Its covariance kernel is given by

$$\mathbf{E}[Z(t)Z(s)] = a^{-2H}R_H(at, as) = R_H(t, s).$$

Since a covariance kernel determines the distribution of a Gaussian process, Z and B_H have the same law.   □

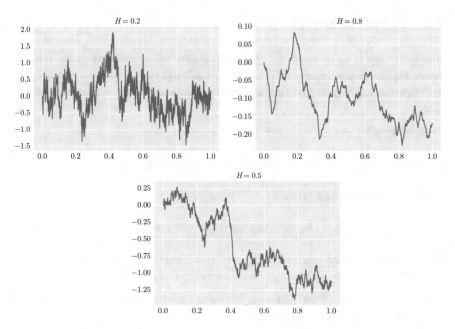

Fig. 4.1 Sample-path example for $H = 0.2$ (upper left), $H = 0.5$ (below), and $H = 0.8$ (upper right)

Theorem 4.2 *With probability* 1, *we have*

$$\lim_{n \to \infty} \sum_{j=1}^{n} \left| B_H\left(\frac{j}{n}\right) - B_H\left(\frac{j-1}{n}\right) \right|^2 = \begin{cases} 0 & \text{if } H > 1/2 \\ \infty & \text{if } H < 1/2. \end{cases}$$

Proof Lemma 4.1 entails that

$$\sum_{j=1}^{n} \left| B_H\left(\frac{j}{n}\right) - B_H\left(\frac{j-1}{n}\right) \right|^{1/H}$$

has the same distribution as

$$\frac{1}{n} \sum_{j=1}^{n} \left| B_H(j) - B_H(j-1) \right|^{1/H}.$$

The ergodic theorem entails that this converges in $L^1(W \to \mathbf{R}; \mu_H)$ and almost-surely to $\mathbf{E}\left[|B_H(1)|^H \right]$, hence the result. □

As a consequence, B_H cannot be a semi-martingale as its quadratic variation is either null or infinite.

4.2   Cameron–Martin Space

The next step is to describe the Cameron–Martin space attached to the fBm of index H. The general theory of Gaussian processes says that we must consider the self-reproducing Hilbert space defined by the covariance kernel, see the appendix of Chap. 1.

Definition 4.2 Let

$$\mathcal{H}^0 = \text{span}\{R_H(t, .), \ t \in [0, 1]\},$$

equipped with the scalar product

$$\langle R_H(t, .), \ R_H(s, .)\rangle_{\mathcal{H}^0} = R_H(t, s). \tag{4.1}$$

The Cameron–Martin space of the fBm of Hurst index H, denoted by $\mathcal{H}_H$, is the completion of $\mathcal{H}^0$ for the scalar product defined in (4.1).

This is not a very practical definition, but we can have a much better description of $\mathcal{H}_H$ thanks to the next theorems.

Lemma 4.2 (Representation of the RKHS) *Assume that there exists a function* $K_H : [0, 1] \times [0, 1] \to \mathbf{R}$ *such that*

$$R_H(s, t) = \int_{[0, 1]} K_H(s, r) \, K_H(t, r) \, \mathrm{d}r \tag{4.2}$$

and that the linear map defined by K_H is one-to-one on $L^2([0, 1] \to \mathbf{R}; \ell)$:

$$\left(\forall t \in [0, 1], \int_{[0, 1]} K_H(t, s) g(s) \, \mathrm{d}s = 0\right) \Longrightarrow g = 0 \ \ell - a.s. \tag{4.3}$$

Then the Hilbert space $\mathcal{H}_H$ can be identified to $K_H\big(L^2([0, 1] \to \mathbf{R}; \ell)\big)$: The space of functions of the form

$$f(t) = \int_{[0, 1]} K_H(t, s) \dot{f}(s) \, \mathrm{d}s$$

for some $\dot{f} \in L^2([0, 1] \to \mathbf{R}; \ell)$, equipped with the inner product

$$\langle K_H f, \ K_H g\rangle_{K_H(L^2([0, 1] \to \mathbf{R}; \ell))} = \langle f, g\rangle_{L^2([0, 1] \to \mathbf{R}; \ell)}.$$

Note that we abused the notations by denoting $K_H^{-1}(f)$ as $\dot{f}$. We will be rewarded of this little infringement below as all the formulas will look the same whatever the value of H.

Proof **Step 1** Equation (4.3) means that

$$\mathfrak{K}_H = \text{span}\{K_H(t, .), \ t \in [0, 1]\}$$

is dense in $L^2([0, 1] \to \mathbf{R}; \ell)$.

Step 2 Since $K_H(K_H(t, .))(s) = R_H(t, s)$,

$$K_H\left(\sum_{k=1}^n \alpha_k K_H(t_k, .)\right) = \sum_{k=1}^n \alpha_k R_H(t_k, .).$$

On the one hand, we have

$$\left\|\sum_{k=1}^n \alpha_k R_H(t_k, .)\right\|_{\mathcal{H}_H}^2 = \sum_{k=1}^n \sum_{l=1}^n \alpha_k \alpha_l R_H(t_k, t_l) \tag{4.4}$$

and on the other hand, we observe that

$$\left\|\sum_{k=1}^n \alpha_k K_H(t_k, .)\right\|_{L^2([0, 1] \to \mathbf{R}; \ell)}^2$$

$$= \int_{[0, 1]} \left(\sum_{k=1}^n \alpha_k K_H(t_k, s)\right)^2 ds$$

$$= \sum_{k=1}^n \sum_{l=1}^n \alpha_k \alpha_l \iint_{[0, 1] \times [0, 1]} K_H(t_k, s) K_H(t_l, s) \, ds \tag{4.5}$$

$$= \sum_{k=1}^n \sum_{l=1}^n \alpha_k \alpha_l R_H(t_k, t_l),$$

in view of (4.2).

Step 3 Equations (4.4) and (4.5) mean that the map K_H:

$$K_H : \mathfrak{K}_H \longrightarrow \mathcal{H}^0$$

$$K_H(t, .) \longrightarrow R_H(t, .)$$

is a bijective isometry when these spaces are equipped with the topology of $L^2([0, 1] \to \mathbf{R}; \ell)$ and $\mathcal{H}^0$, respectively. By density, K_H is a bijective isometry from $L^2([0, 1] \to \mathbf{R}; \ell)$ into $\mathcal{H}_H$. Otherwise stated, $K_H(L^2([0, 1] \to \mathbf{R}; \ell))$ is isometrically isomorphic, hence identified, to $\mathcal{H}_H$.

$\square$

Example (RKHS of the Brownian Motion) For $H = 1/2$, we have

$$t \wedge s = \int_0^1 \mathbf{1}_{[0,t]}(r)\mathbf{1}_{[0,s]}(r)\, dr.$$

This means that the RKHS of the Brownian motion is equal to $I_{1,2}$ since for $K_{1/2}(t,r) = \mathbf{1}_{[0,t]}(r)$,

$$K_{1/2}f(t) = \int_0^1 \mathbf{1}_{[0,t]}(r)\, f(r)\, dr = I^1 f(t).$$

We now have to identify K_H for our kernel R_H.

Lemma 4.3 *For $H > 1/2$, Eq. (4.2) is satisfied with*

$$K_H(t,r) = \frac{r^{1/2-H}}{\Gamma(H-1/2)} \int_r^t u^{H-1/2}(u-r)^{H-3/2}\, du\, \mathbf{1}_{[0,t]}(r). \tag{4.6}$$

Proof According to the fundamental theorem of calculus, applied twice, we can write:

$$R_H(s,t) = \frac{V_H}{4H(2H-1)} \int_0^t \int_0^s |r-u|^{2H-2}\, du\, dr \tag{4.7}$$

After a deep inspection of the handbooks of integrals or more simply, finding, with a bit of luck, the reference [2], we see that

$$\frac{V_H}{4H(2H-1)}|r-u|^{2H-2}$$
$$= (ru)^{H-1/2} \int_0^{r \wedge u} v^{1/2-H}(r-v)^{H-3/2}(u-v)^{H-3/2}\, dv. \tag{4.8}$$

Plug (4.8) into (4.7) and apply Fubini to put the integration with respect to v in the outer most integral. This implies that (4.2) is satisfied with K_H given by (4.6). □

Unfortunately, this integral is not defined for $H < 1/2$ because of the term $(u-r)^{H-3/2}$. Fortunately, the expression (4.6) can be expressed as a hypergeometric function. These somehow classical functions can be presented in different manners so that they are meaningful for a very wide range of parameters, including the domain that is of interest for us.

Definition 4.3 The Gauss hypergeometric function $F(a,b,c,z)$ (for details, see [6]) is defined for any a, b, any z, $|z| < 1$, and any $c \neq 0, -1, \ldots$ by

$$F(a,b,c,z) = \sum_{k=0}^{+\infty} \frac{(a)_k (b)_k}{(c)_k k!} z^k, \tag{4.9}$$

where $(a)_0 = 1$ and $(a)_k = \Gamma(a+k)/\Gamma(a) = a(a+1)\ldots(a+k-1)$ is the Pochhammer symbol.

If a or b is a negative integer, the series terminates after a finite number of terms and $F(a, b, c, z)$ is a polynomial in z.

The radius of convergence of this series is 1, and there exists a finite limit when z tends to 1 ($z < 1$) provided that $\Re(c - a - b) > 0$.

For any z such that $|\arg(1 - z)| < \pi$, any a, b, c such that $\Re(c) > \Re(b) > 0$, F can be defined by

$$F(a, b, c, z) = \frac{\Gamma(c)}{\Gamma(b)\Gamma(c - b)} \int_0^1 u^{b-1}(1 - u)^{c-b-1}(1 - zu)^{-a} \, du. \qquad (4.10)$$

Remark 4.1 The Gamma function is defined by an integral only on $\{z, \Re(z) > 0\}$. By the famous relation, $\Gamma(z+1) = z\Gamma(z)$, it can be extended analytically to $\mathbf{C}\backslash(-\mathbf{N})$ even if the integral expression is no longer valid. The same but more involved kind of reasoning can be done here to extend F.

Theorem 4.3 *The hypergeometric function F can be extended analytically to the domain $\mathbf{C} \times \mathbf{C} \times \mathbf{C}\backslash(-\mathbf{N}) \times \{z, |\arg(1 - z)| < \pi\}$.*

Proof We will not go into the details of the proof. Given (a, b, c), consider Σ the set of triples (a', b', c') such that $|a - a'| = 1$ or $|b - b'| = 1$ or $|c - c'| = 1$. Any hypergeometric function $F(a', b', c', z)$ with (a', b', c') in Σ is said to be contiguous to $F(a, b, c)$. For any two hypergeometric functions F_1 and F_2 contiguous to $F(a, b, c, z)$, there exists a relation of the type:

$$P_0(z)F(a, b, c, z) + P_1(z)F_1(z) + P_2(z)F_2(z) = 0, \text{ for } z, \ |\arg(1 - z)| < \pi,$$
$$(4.11)$$

where for any i, P_i is a polynomial with respect to z. These relations permit to define the analytic continuation of $F(a, b, c, z)$ with respect to its four variables. $\qquad \square$

If we want to have a representation similar to (4.6) for $H < 1/2$, we need to write K in a form that can be extended to larger domain. The easiest way to proceed is to write K as an entire function of its arguments H, t and r. That is where hypergeometric function enters the scene.

Theorem 4.4 *For any $H \in (0, 1)$, R_H can be factorized as in (4.2) with*

$$K_H : [0, 1]^2 \longrightarrow \mathbf{R}$$

$$(t, s) \longmapsto \frac{(t - s)^{H-1/2}}{\Gamma(H + 1/2)} F\left(H - 1/2, 1/2 - H, H + 1/2, 1 - \frac{t}{s}\right).$$
$$(4.12)$$

If we identify integral operators and their kernel, this amounts to say that

$$R_H = K_H \circ K_H^*.$$

Proof For $H > 1/2$, a change of variable in (4.6) transforms the integral term into

$$(t - r)^{H-1/2} r^{H-1/2} \int_0^1 u^{H-3/2} (1 - (1 - t/r)u)^{H-1/2} \, du.$$

By the definition (4.10) of hypergeometric functions, we see that (4.12) holds true for $H > 1/2$. According to the properties of the hypergeometric function, we have

$$K_H(t, r) = \frac{2^{-2H} \sqrt{\pi}}{\Gamma(H) \sin(\pi H)} r^{H-1/2}$$
$$+ \frac{1}{2\Gamma(H + 1/2)} (t - r)^{H-1/2} F(1/2 - H, 1, 2 - 2H, \frac{r}{t}).$$

If $H < 1/2$, then the hypergeometric function of the latter equation is continuous with respect to r on $[0, t]$ because $2 - 2H - 1 - 1/2 + H = 1/2 - H$ is positive. Hence, for $H < 1/2$, $K_H(t, r)(t - r)^{1/2-H} r^{1/2-H}$ is continuous with respect to r on $[0, t]$. For $H > 1/2$, the hypergeometric function is no more continuous in t, but we have [6]

$$F(1/2 - H, 1, 2 - 2H, \frac{r}{t}) = C_1 F(1/2 - H, 1, H + 1/2, 1 - \frac{r}{t})$$
$$+ C_2 (1 - \frac{r}{t})^{1/2-H} (\frac{r}{t})^{2H-1}.$$

Hence, for $H \geq 1/2$, $K_H(t, r) r^{H-1/2}$ is continuous with respect to r on $[0, t]$. Fix $\delta \in [0, 1/2)$ and $t \in (0, 1]$, we have

$$|K_H(t, r)| \leq C r^{-|H-1/2|} (t - r)^{-(1/2-H)+} \mathbf{1}_{[0,t]}(r)$$

where C is uniform with respect to $H \in [1/2 - \delta, 1/2 + \delta]$. Thus, the two functions defined on $\{H \in \mathbf{C}, \ |H - 1/2| < 1/2\}$ by

$$H \longmapsto R_H(s, t) \text{ and } H \longmapsto \int_0^1 K_H(s, r) K_H(t, r) \, dr$$

are well defined, analytic with respect to H, and coincide on $[1/2, 1)$; thus they are equal for any $H \in (0, 1)$ and any s and t in $[0, 1]$. □

In the previous proof, we proved a result that is so useful in its own that it deserves to be a theorem:

Theorem 4.5 *For any $H \in (0, 1)$, for any t, the function*

$$[0, t] \longrightarrow \mathbf{R}$$

$$r \longmapsto K_H(t, r) r^{|H-1/2|} (t - r)^{(1/2-H)_+}$$

is continuous on $[0, t]$.

Moreover, there exists a constant c_H such that for any $0 \le r \le t \le 1$

$$|K_H(t, r)| \le c_H \, r^{-|H-1/2|} (t - r)^{-(1/2-H)_+}. \tag{4.13}$$

These continuity results are illustrated by the following pictures.

We made some progress with this new description of $\mathcal{H}_H$. However, for a given element of $L^2([0, 1] \to \mathbf{R}; \ell)$, it is still difficult to determine whether it belongs to $\mathcal{H}_H$. Since

$$\int_0^1 \int_0^1 K(t, r)^2 \, \mathrm{d}t \, \mathrm{d}r = \int_0^1 R_H(t, t) \, \mathrm{d}t < \infty,$$

we already know that the integral map of kernel K_H is Hilbert–Schmidt from $L^2([0, 1] \to \mathbf{R}; \ell)$ into itself. Thanks to [8, page 187], we are in position to give a fully satisfactory description of $\mathcal{H}_H$ (Fig. 4.2).

Theorem 4.6 *Consider the integral transform of kernel K_H, i.e.,*

$$K_H : L^2([0, 1] \to \mathbf{R}; \ell) \longrightarrow L^2([0, 1] \to \mathbf{R}; \ell)$$

$$f \longmapsto \left(t \mapsto \int_0^t K_H(t, s) f(s) \, \mathrm{d}s \right).$$

The map K_H is an isomorphism from $L^2([0, 1] \to \mathbf{R}; \ell)$ onto $I_{H+1/2,2}$ (see Definition 1.7), and we have the following representations, which says that K_H is in some sense, close to the map $I^{H+1/2}$:

$$K_H f = I_{0^+}^{2H} x^{1/2-H} I_{0^+}^{1/2-H} x^{H-1/2} f \text{ for } H \le 1/2,$$

$$K_H f = I_{0^+}^{1} x^{H-1/2} I_{0^+}^{H-1/2} x^{1/2-H} f \text{ for } H \ge 1/2.$$

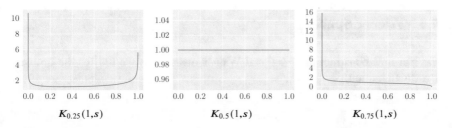

$K_{0.25}(1, s)$ $K_{0.5}(1, s)$ $K_{0.75}(1, s)$

Fig. 4.2 $K_H(1, .)$ for $H = 0.2$, $H = 0.8$, and $H = 0.5$

Note that if $H \geq 1/2$, $r \to K_H(t, r)$ is continuous on $(0, t]$ so that we can include t in the indicator function.

Remark 4.2 We already know that the fBm is all the more regular than its Hurst index is close to 1. However, we see that the kernel K_H is more and more singular when H goes to 1. This means that it is probably a bad idea to devise properties of B_H using the properties of K_H. On the other hand, as an operator, K_H is more and more regular as H increases. This indicates that the efficient approach is to work with K_H as an operator. We tried to illustrate this line of reasoning in the next results.

To summarize the previous considerations, we get:

Theorem 4.7 (RKHS of the fBm) *The Cameron–Martin of the fractional Brownian motion is $\mathcal{H}_H = \{K_H \dot{h}; \ \dot{h} \in L^2([0, 1] \to \mathbf{R}; \ell)\}$, i.e., any $h \in \mathcal{H}_H$ can be represented as*

$$h(t) = K_H \dot{h}(t) = \int_0^1 K_H(t, s) \dot{h}(s) \, ds,$$

where $\dot{h}$ belongs to $L^2([0, 1] \to \mathbf{R}; \ell)$. For any $\mathcal{H}_H$-valued random variable u, we hereafter denote by $\dot{u}$ the $L^2([0, 1] \to \mathbf{R}; \ell)$-valued random variable such that

$$u(w, t) = \int_0^t K_H(t, s) \dot{u}(w, s) \, ds.$$

The scalar product on $\mathcal{H}_H$ is given by

$$(h, g)_{\mathcal{H}_H} = (K_H \dot{h}, K_H \dot{g})_{\mathcal{H}_H} = (\dot{h}, \dot{g})_{L^2([0,1] \to \mathbf{R}; \ell)}.$$

Remark 4.3 Theorem 4.6 implies that as a vector space, $\mathcal{H}_H$ is equal to $I_{H+1/2,2}$, but the norm on each of these spaces is different since

$$\|K_H \dot{h}\|_{\mathcal{H}_H} = \|\dot{h}\|_{L^2([0,1] \to \mathbf{R}; \ell)}$$

$$\text{and } \|K_H \dot{h}\|_{I_{H+1/2,2}} = \|(I_{0+}^{-H-1/2} \circ K_H) \dot{h}\|_{L^2([0,1] \to \mathbf{R}; \ell)}.$$

4.3 Wiener Space

We can now construct the fractional Wiener measure as we did for the ordinary Brownian motion.

Theorem 4.8 *Let* $(\dot{h}_m, m \geq 0)$ *be a complete orthonormal basis of* $L^2([0, 1] \to$ $\mathbf{R}; \ell)$ *and* $h_m = K_H \dot{h}_m$. *Consider the sequence*

$$S_n^H(t) = \sum_{m=0}^{n} X_m h_m(t)$$

where $(X_m, m \geq 0)$ *is a sequence of independent standard Gaussian random variables. Then,* $(S_n^H, n \geq 0)$ *converges, with probability* 1, *in* $W_{\alpha,p}$ *for any* $\alpha < H$ *and any* $p > 1$.

Proof The proof proceeds exactly as the proof of Theorem 1.5. The trick is to note that

$$(h_m(t) - h_m(s))^2 = \langle K_H(t, .) - K_H(s, .), \dot{h}_m \rangle_{\mathcal{H}_H}^2,$$

so that

$$\sum_{m=0}^{\infty} (h_m(t) - h_m(s))^2 = \|K_H(t, .) - K_H(s, .)\|_{L^2([0,1] \to \mathbf{R}; \ell)}^2$$

$$= R_H(t, t) - R_H(s, s) - 2R_H(t, s) = V_H|t - s|^{2H}.$$

Moreover,

$$\int_{[0,1]^2} |t - s|^{pH-1-\alpha p} \, ds \, dt < \infty \text{ if and only if } \alpha < H.$$

This means, by dominated convergence, that

$$\sup_{n \geq M} \mathbf{E}\left[\|S_n^H - S_M^H\|_{W_{\alpha,p}}^p\right]$$

$$\leq c \iint_{[0,1]^2} \Big(\sum_{m=M+1}^{\infty} (h_m(t) - h_m(s))^2\Big)^{p/2} |t - s|^{-1-\alpha p} \, ds \, dt \xrightarrow{M \to \infty} 0,$$

provided that $\alpha < H$. The proof is finished as in Theorem 1.5. $\square$

   In what follows, W may be taken either as $C_0([0, 1], \mathbf{R})$ or as any of the spaces $W_{\gamma,p}$ with

$$p \geq 1, \; 1/p < \gamma < H.$$

Fig. 4.3 Embeddings and
identification for the Gelfand
triplet in the fBm

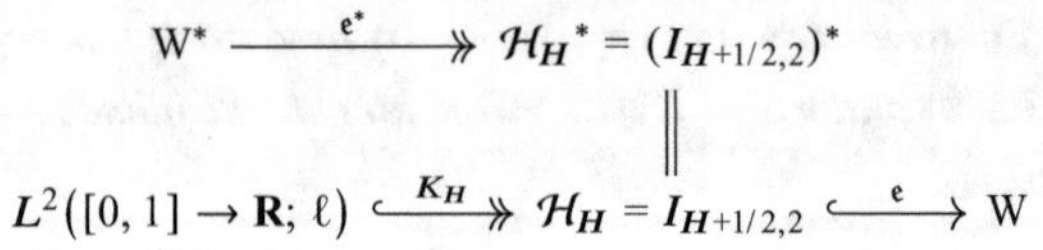

$$\text{W}^* \xrightarrow{\ e^*\ } \mathcal{H}_H{}^* = (I_{H+1/2,2})^*$$

$$\Big\|$$

$$L^2([0,1] \to \mathbf{R};\ \ell) \xhookrightarrow{\ K_H\ } \mathcal{H}_H = I_{H+1/2,2} \xhookrightarrow{\ e\ } \text{W}$$

For any $H \in (0, 1)$, μ_H is the unique probability measure on W such that the canonical process $(B_H(s);\ s \in [0, 1])$ is a centered Gaussian process with covariance kernel R_H:

$$\mathbf{E}\,[B_H(s)B_H(t)] = R_H(s, t).$$

The canonical filtration is given by $\mathcal{F}_t^H = \sigma\{B_H(s),\ s \le t\} \vee \mathcal{N}_H$, and $\mathcal{N}_H$ is the set of the μ_H-negligible events. The analog of the diagram of Fig. 1.2 reads as the diagram of Fig. 4.3.

We can, as before, search for the image of ε_t by e^*. We have, for $h \in \mathcal{H}_H$, on the one hand,

$$h(t) = \langle \varepsilon_t, e(h) \rangle_{\text{W}^*,\text{W}} = \langle e^*(\varepsilon_t),\ h \rangle_{\mathcal{H}_H}.$$

On the other hand,

$$h(t) = K_H \dot{h}(t) = \langle K_H(t, .),\ \dot{h} \rangle_{L^2\left([0,1] \to \mathbf{R};\ \ell\right)} = \langle R_H(t, .),\ h \rangle_{\mathcal{H}_H}.$$

Hence,

$$e^*(\varepsilon_t) = R_H(t, .) \text{ and } K_H^{-1}(e^*(\varepsilon_t)) = K_H(t, .).$$

Recall that for the ordinary Brownian motion, we have

$$e^*(\varepsilon_t) = t \wedge . = R_{1/2}(t, .) \text{ and } K_{1/2}^{-1}(e^*(\varepsilon_t)) = \mathbf{1}_{[0,t]}(.) = K_{1/2}(t, .).$$

Theorem 4.9 *For any z in W^*,*

$$\int_{\text{W}} e^{i\langle z, \omega \rangle_{\text{W}^*,\text{W}}}\, d\mu_H(\omega) = \exp\left(-\frac{1}{2}\|e^*(z)\|_{\mathcal{H}_H}^2\right). \tag{4.14}$$

Proof By dominated convergence, we have

$$\int_{\text{W}} e^{i\langle z, \omega \rangle_{\text{W}^*,\text{W}}}\, d\mu_H(\omega) = \lim_{n \to \infty} \mathbf{E}\left[\exp\left(i \sum_{m=0}^{n} X_m \langle z,\ e(K_H \dot{h}_m) \rangle_{\text{W}^*,\text{W}}\right)\right]$$

$$= \lim_{n \to \infty} \exp\left(-\frac{1}{2} \sum_{m=0}^{n} \langle e^*(z),\ K_H \dot{h}_m \rangle_{\mathcal{H}_H}^2\right)$$

$$= \exp\left(-\frac{1}{2}\sum_{m=0}^{\infty}\langle e^*(z), K_H\dot{h}_m\rangle_{\mathcal{H}_H}^2\right)$$

$$= \exp\left(-\frac{1}{2}\|e^*(z)\|_{\mathcal{H}_H}^2\right),$$

according to the Parseval identity. □

The Wiener integral is constructed as before as the extension of the map

$$\delta_H : W^* \subset I_{1,2} \longrightarrow L^2(\mu_H)$$

$$z \longmapsto \langle z, B_H\rangle_{W^*,W}.$$

By construction of the Wiener measure, the random variable $\langle z, B_H\rangle_{W^*,W}$ is Gaussian with mean 0 and variance $\|R_H(z)\|_{\mathcal{H}_H}^2$. For $z = \varepsilon_t$, we have

$$B_H(t) = \langle \varepsilon_t, B_H\rangle_{W^*,W} = \delta_H(R_H(t, .)).$$

Equation (4.14) is the exact analog of Eq. (1.13); hence, the Cameron–Martin theorem can be proved identically:

Theorem 4.10 *For any $h \in \mathcal{H}_H$, for any bounded $F : W \to \mathbf{R}$,*

$$\mathbf{E}[F(B_H + e(h)] = \mathbf{E}\left[F(B_H)\exp\left(\delta_H(h) - \frac{1}{2}\|h\|_{\mathcal{H}_H}^2\right)\right]. \tag{4.15}$$

For the Brownian motion, it is often easier to work with elements of $L^2([0, 1] \to \mathbf{R}; \ell)$ instead of their image by $K_{1/2}$, which belongs to $I_{1,2}$. If we try to mimic this approach for the fractional Brownian motion, we should write

$$B_H(t) = \delta_H(R_H(t, .)) = \delta_H(K_H(K_H(t, .))) = \int_0^1 K_H(t, s)\,\delta B_H(s),$$

which has to be compared to

$$B(t) = B_{1/2}(t) = \int_0^1 \mathbf{1}_{[0,t]}(s)\,dB_{1/2}(s),$$

where the integral is taken in the Itô sense. Remark that these two equations are coherent since $K_{1/2}(t, .) = \mathbf{1}_{[0,t]}$.

Lemma 4.4 *The process $B = \left(\delta_H(K_H(\mathbf{1}_{[0,t]})), t \in [0, 1]\right)$ is a standard Brownian motion. For $u \in L^2([0, 1] \to \mathbf{R}; \ell)$,*

$$\int_0^1 u(s)\,dB(s) = \delta_H(K_H u). \tag{4.16}$$

In particular,

$$B_H(t) = \int_0^t K_H(t, s) \, dB(s). \tag{4.17}$$

Proof It is a Gaussian process by the definition of the Wiener integral. We just have to verify that it has the correct covariance kernel: It suffices to see that $\|K_H(\mathbf{1}_{[0,t]})\|_{\mathcal{H}_H}^2 = t$. But,

$$\|K_H(\mathbf{1}_{[0,t]})\|_{\mathcal{H}_H}^2 = \|\mathbf{1}_{[0,t]}\|_{L^2([0,1]\to\mathbf{R};\,\ell)}^2 = t.$$

This means that (4.16) holds for $u = \mathbf{1}_{[0,t]}$; hence, for all piecewise constant functions u and by density, for all $u \in L^2([0,1] \to \mathbf{R};\,\ell)$. $\qquad\square$

Remark 4.4 Equation (4.17) is known as the Karhunen–Loeve representation. We could have started by considering a process defined by the right-hand side of (4.17) and called it fractional Brownian motion. Actually, (4.17) is a stronger result: It says that starting from an fBm, one can construct a Brownian motion on the same probability space such that the representation (4.17) holds.

The following theorem is an easy consequence of the properties of the maps K_H.

Theorem 4.11 *The operator* $\mathcal{K}_H = K_H \circ K_{1/2}^{-1}$ *is continuous and invertible from* $I_{\alpha,p}$ *into* $W_{\alpha+H-1/2,p}$, *for any* $\alpha > 0$.

Formally, we have $B_H = K_H(\dot{B}) = K_H \circ K_{1/2}^{-1}(B)$, so we can expect that:

Theorem 4.12 (B Is a Function of B_H) *For any H, we have*

$$B_H \overset{dist}{=} \mathcal{K}_H(B) \quad \text{and} \quad B \overset{dist}{=} \mathcal{K}_H^{-1}(B_H) \tag{4.18}$$

Proof Let $(\dot{h}_m, m \geq 0)$ be a complete orthonormal basis of $L^2([0,1] \to \mathbf{R};\,\ell)$. The series, which defines B,

$$B = \sum_{m=0}^{\infty} X_m I^1(\dot{h}_m),$$

converges with $\mu_{1/2}$-probability 1, in any $W_{\alpha,p}$, provided that $0 < \alpha - 1/p < 1/2$. By continuity of $\mathcal{K}_H$,

$$\mathcal{K}_H\left(\sum_{m=0}^{\infty} X_m I^1(\dot{h}_m)\right) = \sum_{m=0}^{\infty} X_m K_H(\dot{h}_m) \overset{dist}{=} B_H$$

converges on the same set of full measure in $I_{\alpha+H-1/2,p}$. Note that when $\alpha - 1/p$ runs through $(0, 1/2)$, $\alpha + H - 1/2 - 1/p$ varies along $(0, H)$. Hence, we retrieve the desired regularity of the sample-paths of B_H.

The same proof holds for the second identity. $\qquad\qquad\qquad\qquad\qquad\square$

4.4 Gradient and Divergence

The gradient is defined as for the usual Brownian motion. The only modification is the Cameron–Martin space.

Definition 4.4 A function F is said to be cylindrical if there exists an integer n, $f \in$ Schwartz$(\mathbf{R}^n)$, the Schwartz space on $\mathbf{R}^n$, $(h_1, \cdots, h_n)$, n elements of $\mathcal{H}_H$ such that

$$F(\omega) = f\left(\delta_H h_1(\omega), \cdots, \delta_H h_n(\omega)\right).$$

The set of such functionals is denoted by $\mathcal{S}_{\mathcal{H}_H}$.

Definition 4.5 Let $F \in \mathcal{S}_{\mathcal{H}_H}, h \in \mathcal{H}_H$, with $F(\omega) = f\left(\delta_H h_1(\omega), \cdots, \delta_H h_n(\omega)\right)$. Set

$$\nabla F = \sum_{j=1}^{n} \partial_j f\left(\delta_H h_1, \cdots, \delta_H h_n\right) h_j,$$

so that

$$\langle \nabla F, h \rangle_{\mathcal{H}_H} = \sum_{j=1}^{n} \partial_j f\left(\delta_H h_1, \cdots, \delta_H h_n\right) \langle h_j, h \rangle_{\mathcal{H}_H}.$$

Example (Derivative of $f(B_H(t))$) This means that

$$\nabla f\left(B_H(t)\right) = f'\left(B_H(t)\right) R_H(t, .)$$

and if we denote $\dot{\nabla} = K_H^{-1} \nabla$ (which corresponds for $H = 1/2$ to take the time derivative of the gradient), we get

$$\dot{\nabla}_s f\left(B_H(t)\right) = f'\left(B_H(t)\right) K_H(t, s).$$

We can now improve Theorem 4.12.

Theorem 4.13 *Let*

$$B_H(t) = \delta_H\big(R_H(t, .)\big) \text{ and } B(t) = \delta_H\big(K_H(\mathbf{1}_{[0,t]})\big).$$

For any H, we have

$$\mu_H\Big(B = \mathcal{K}_H^{-1}(B_H)\Big) = 1. \tag{4.19}$$

Integrate by Parts You Shall

When facing stochastic integrals or divergence, it is always a good idea to proceed to as many integration by parts as necessary to obtain ordinary integrals with respect to the Lebesgue measure. Then, we can modify them by the usual tools (dominated convergence, Fubini, etc.) and redo the integration by parts. This is the general scheme of the following proof and of several others as the Itô formula.

***Proof* Step 1** The sample-paths of B are known to be continuous, and those of B_H belong to $W_{H-\varepsilon,p}$ for any $p \geq 1$ and ε sufficiently small. Hence, according to Theorem 4.11, $\mathcal{K}_H^{-1}(B_H)$ almost-surely belongs to $I_{1/2-\varepsilon,p}$ for any $p \geq 1$. Choose $p > 2$ so that $I_{1/2-\varepsilon,p} \subset C_0([0, 1], \mathbf{R})$ to conclude that $\mathcal{K}_H^{-1}(B_H)$ has μ_H-a.s. continuous sample-paths.

Step 2 To prove such an identity, it is necessary and sufficient to check that

$$\mathbf{E}\left[\psi \int_0^1 B(t)g(t)\,\mathrm{d}t\right] = \mathbf{E}\left[\psi \int_0^1 \mathcal{K}_H^{-1}(B_H)(t)\,g(t)\,\mathrm{d}t\right] \tag{4.20}$$

for any $g \in L^2\big([0, 1] \to \mathbf{R};\ \ell\big)$ and any $\psi \in \mathcal{S}_H$. Indeed, $L^2\big([0, 1] \to \mathbf{R};\ \ell\big) \otimes \mathcal{S}_H$ is a dense subset of $L^2\big([0, 1] \to \mathbf{R};\ \ell\big) \otimes L^2\big(\mathbf{W} \to \mathbf{R};\ \mu_H\big) \simeq L^2\big([0, 1] \otimes \mathbf{W} \to \mathbf{R};\ \ell \otimes \mu_H\big)$, and (4.20) entails that $B = \mathcal{K}_H^{-1}(B_H)$ $\ell \otimes \mu_H$-almost-surely. This means that there exists $A \subset [0, 1] \times \mathbf{W}$ such that

$$\int_{[0,1]\times W} \mathbf{1}_A(s, \omega)\,\mathrm{d}s\,\mathrm{d}\mu_H(\omega) = 0,$$

and

$$B(\omega, s) = \mathcal{K}_H^{-1}(B_H)(\omega, s) \text{ for } (s, \omega) \notin A.$$

Hence, for any $s \in [0, 1]$, the section of A at s fixed, i.e., $A_s = \{\omega, (s, \omega) \in A\}$, is a μ_H-negligeable set. Now, consider

$$A_Q = \bigcup_{t \in [0,1] \cap Q} A_t.$$

It is a μ_H-negligeable set, and for $\omega \in A_Q^c$, for $t \in [0, 1] \cap Q$, $B(\omega, s) = \mathcal{K}_H^{-1}(B_H)(\omega, s)$. Thus, by continuity, this identity still holds for any $t \in [0, 1]$ and any $\omega \in A_Q^c$. This means that Eq. (4.19) holds.

Step 3 We now prove (4.20),

$$\mathbf{E}\left[\psi \int_0^1 \mathcal{K}_H^{-1}(B_H)(t)\, g(t)\, dt\right] = \int_0^1 \mathbf{E}\left[\psi\, B_H(t)\right] (\mathcal{K}_H^{-1})^*(g)(t)\, dt$$

$$= \int_0^1 \mathbf{E}\left[\psi\, \delta_H(R_H(t, .))\right] (\mathcal{K}_H^{-1})^*(g)(t)\, dt$$

$$= \mathbf{E}\left[\int_0^1 (\mathcal{K}_H^{-1})^*(g)(t) \int_0^1 \dot{\nabla}_s \psi\, K_H(t, s)\, ds\, dt\right]$$

$$= \mathbf{E}\left[\int_0^1 \dot{\nabla}_s \psi \int_0^1 K_H(t, s)(\mathcal{K}_H^{-1})^*(g)(t)\, dt\, ds\right]$$

$$= \mathbf{E}\left[\int_0^1 \dot{\nabla}_s \psi\, K_H^*(\mathcal{K}_H^{-1})^*(g)(s)\, ds\right]$$

By the very definition of $\mathcal{K}_H$,

$$K_H^* \circ (\mathcal{K}_H^{-1})^* = K_H^* \circ (K_H^{-1})^* \circ K_{1/2}^* = K_{1/2}^*.$$

Thus, we have

$$\mathbf{E}\left[\psi \int_0^1 \mathcal{K}_H^{-1}(B_H)(t)\, g(t)\, dt\right] = \mathbf{E}\left[\int_0^1 \dot{\nabla}_s \psi\, K_{1/2}^* g(s)\, ds\right]$$

$$= \mathbf{E}\left[\int_0^1 \dot{\nabla}_s \psi \int_s^1 g(t)\, dt\, ds\right]$$

$$= \mathbf{E}\left[\int_0^1 \int_0^1 \dot{\nabla}_s \psi\, g(t)\, \mathbf{1}_{[s,1]}(t)\, dt\, ds\right]$$

$$= \mathbf{E}\left[\int_0^1 \int_0^1 \dot{\nabla}_s \psi\, g(t)\, \mathbf{1}_{[0,t]}(s)\, dt\, ds\right]$$

$$= \mathbf{E}\left[\int_0^1 g(t) \int_0^1 \dot{\nabla}_s \psi\, \mathbf{1}_{[0,t]}(s)\, ds\, dt\right].$$

On the other hand, $B(t) = \delta_H\big(K_H(\mathbf{1}_{[0,t]})\big)$; hence,

$$\mathbf{E}\left[\psi \int_0^1 B(t)g(t)\,dt\right] = \mathbf{E}\left[\psi \int_0^1 \delta_H\big(K_H(\mathbf{1}_{[0,t]})\big)\,g(t)\,dt\right]$$

$$= \mathbf{E}\left[\int_0^1 g(t) \int_0^1 \dot{\nabla}_s \psi \, \mathbf{1}_{[0,t]}(s)\,ds\,dt\right].$$

Then, (4.20) follows.

$\square$

We can even go further and show that B and B_H generate the same filtration.

Definition 4.6 Recall that $(\dot{\pi}_t, t \in [0,1])$ are the projections defined by

$$\dot{\pi}_t : L^2\big([0,1] \to \mathbf{R}; \ell\big) \longrightarrow L^2\big([0,1] \to \mathbf{R}; \ell\big)$$

$$f \longmapsto f\mathbf{1}_{[0,t)}.$$

Let V be a closable map from Dom $V \subset L^2([0,1] \to \mathbf{R}; \ell)$ into $L^2([0,1] \to \mathbf{R}; \ell)$. Then, V is $\dot{\pi}$-causal if Dom V is $\dot{\pi}$-stable, i.e., $\dot{\pi}_t$ Dom $V \subset$ Dom V for any $t \in [0,1]$, and if for any $t \in [0,1]$,

$$\dot{\pi}_t V \dot{\pi}_t = \dot{\pi}_t V.$$

Consider also π_t^H defined by

$$\pi_t^H : \mathcal{H}_H \longrightarrow \mathcal{H}_H$$

$$h \longmapsto K_H\big(\pi_t K_H^{-1}(h)\big) = K_H\big(\dot{h}\,\mathbf{1}_{[0,t]}\big).$$

Remark 4.5 An integral operator, i.e.,

$$Vf(t) = \int_0^1 V(t,s)f(s)\,ds,$$

is $\dot{\pi}$-causal if and only if $V(t,s) = 0$ for $s > t$. For V_1, V_2 two causal operators, their composition $V_1 V_2$ is still causal:

$$\pi_t V_1 V_2 \pi_t = (\pi_t V_1 \pi_t) V_2 \pi_t = \pi_t V_1 (\pi_t V_2 \pi_t)$$

$$= \pi_t V_1 (\pi_t V_2) = (\pi_t V_1 \pi_t) V_2 = \pi_t V_1 V_2.$$

Corollary 4.1 *The filtrations generated by B_H and B do coincide.*

Proof From the representation

$$B_H(t) = \int_0^t K_H(t, s) \, dB(s),$$

we deduce that

$$\sigma \{B_H(s), \ s \le t\} \subset \sigma \{B(s), \ s \le t\}.$$

We have $\mathcal{K}_H^{-1} = K_{1/2} K_H^{-1}$. From Theorem 4.6, K_H^{-1} appears as the composition of fractional derivatives and multiplication operators:

$$f \mapsto x^\alpha f.$$

Time derivatives of any order (as in Definition 4.11) are clearly causal operators. It is straightforward that multiplication operators are also causal. Thus, $\mathcal{K}_H^{-1}$ appears as the composition of causal operators; hence, it is causal. In view of (4.19), this means that

$$B(t) = \int_0^t V(t, s) B_H(s) \, ds$$

for some lower triangular kernel V. Hence,

$$\sigma \{B_H(s), \ s \le t\} \supset \sigma \{B(s), \ s \le t\},$$

and the equality of filtrations is proved. □

We can now reap the fruits of our not so usual presentation of the Malliavin calculus for the Brownian motion, in which we cautiously sidestepped chaos decomposition. Theorem 4.10 entails the integration by parts formula, pending of (2.5): For any F and G in $\mathcal{S}_H$, for any $h \in \mathcal{H}_H$,

$$\mathbf{E}\left[G \langle \nabla F, h \rangle_{\mathcal{H}_H}\right] = -\mathbf{E}\left[F \langle \nabla G, h \rangle_{\mathcal{H}_H}\right] + \mathbf{E}[FG \, \delta_H h]. \tag{4.21}$$

Definition 4.5 is formally the very same as Definition 2.2 so that the definitions of the Sobolev spaces are identical.

Definition 4.7 The space $\mathbb{D}_{p,1}^H$ is the closure of $\mathcal{S}_H$ for the norm

$$\|F\|_{p,1,H} = \mathbf{E}\left[|F|^p\right]^{1/p} + \mathbf{E}\left[\|\nabla F\|_{\mathcal{H}_H}^p\right]^{1/p}.$$

The iterated gradients are defined likewise and so do the Sobolev of higher order, $\mathbb{D}_{p,k}^{H}$. We sill clearly have

$$\nabla(FG) = F\nabla G + G\nabla F$$

$$\nabla\phi(F) = \phi'(F)\nabla F$$

for $F \in \mathbb{D}_{p,1}^{H}$, $G \in \mathbb{D}_{q,1}^{H}$, and ϕ Lipschitz continuous. As long as we do not use the temporal scale, there is no difference between the identities established for the usual Brownian motion and those relative to the fractional Brownian motion.

Theorem 4.14 *For any F in $L^2(W \to \mathbf{R}; \mu_H)$,*

$$\Gamma(\pi_t^H)F = \mathbf{E}\left[F \mid \mathcal{F}_t^H\right],$$

in particular,

$$\mathbf{E}\left[B_H(t) \mid \mathcal{F}_r^H\right] = \int_0^t K_H(t,s)\mathbf{1}_{[0,r]}(s)\,\delta B(s), \ and$$

$$\mathbf{E}\left[\exp(\delta_H u - 1/2\|u\|_{\mathcal{H}_H}^2) \mid \mathcal{F}_t^H\right] = \exp(\delta_H \pi_t^H u - 1/2\|\pi_t^H u\|_{\mathcal{H}_H}^2),$$

for any $u \in \mathcal{H}_H$.

Proof Let $\{h_n, n \geq 0\}$ be a denumerable family of elements of $\mathcal{H}_H$, and let $V_n = \sigma\{\delta_H h_k, 1 \leq k \leq n\}$. Denote by p_n the orthogonal projection on span$\{h_1, \ldots, h_n\}$. For any f bounded, for any $u \in \mathcal{H}_H$, by the Cameron–Martin theorem, we have

$$\mathbf{E}\left[\Lambda_1^u f(\delta_H h_1, \ldots, \delta_H h_n)\right]$$

$$= \mathbf{E}[f(\delta_H h_1(w+u), \ldots, \delta_H h_n(w+u))]$$

$$= \mathbf{E}\left[f(\delta_H h_1 + (h_1, u)_{\mathcal{H}_H}, \ldots, \delta_H h_n + (h_n, u)_{\mathcal{H}_H})\right]$$

$$= \mathbf{E}[f(\delta_H h_1(w + p_n u), \ldots, \delta_H h_n(w + p_n u))]$$

$$= \mathbf{E}\left[\Lambda_1^{p_n u} f(\delta_H h_1, \ldots, \delta_H h_n)\right],$$

hence

$$\mathbf{E}\left[\Lambda_1^u \mid V_n\right] = \Lambda_1^{p_n u}. \tag{4.22}$$

Choose h_n of the form $\pi_t^H(e_n)$ where $\{e_n, n \geq 0\}$ is an orthonormal basis of $\mathcal{H}_H$, i.e., $\{h_n, n \geq 0\}$ is an orthonormal basis of $\pi_t^H(\mathcal{H}_H)$. By the previous theorem,

$\bigvee_n V_n = \mathcal{F}_t^H$, and it is clear that p_n tends pointwise to π_t^H; hence, from (4.22) and martingale convergence theorem, we can conclude that

$$\mathbf{E}\left[\Lambda_1^u \mid \mathcal{F}_t^H\right] = \Lambda_1^{\pi_t^H u} = \Lambda_t^u.$$

Moreover, for $u \in \mathcal{H}_H$,

$$\Gamma(\pi_t^H)(\Lambda_1^u) = \Lambda_1^{\pi_t^H u};$$

hence, by density of linear combinations of Wick exponentials, for any $F \in L^2(\mu_H)$,

$$\Gamma(\pi_t^H)F = \mathbf{E}\left[F \mid \mathcal{F}_t^H\right],$$

and the proof is completed. □

Definition 4.8 For the sake of notations, we set, for $\dot{u}$ such that $K_H \dot{u}$ belongs to $\mathrm{Dom}_p \, \delta_H$ for some $p > 1$,

$$\int_0^1 \dot{u}(s) \delta B(s) = \delta_H(K_H \dot{u}) \quad \text{and} \quad \int_0^t \dot{u}(s) \delta B(s) = \delta_H(\pi_t^H K_H \dot{u}). \tag{4.23}$$

Note that, for any $\psi \in \mathbb{D}_{p/(p-1),1}^H$,

$$\mathbf{E}\left[\psi \int_0^1 \dot{u}(s) \delta B(s)\right] = \mathbf{E}\left[\int_0^1 \dot{\nabla}_s \psi \, \dot{u}(s) \, ds\right].$$

The next result is the Clark formula. It reads formally as (3.15), but we should take care that the $\dot{\nabla}$ does not represent the same object. Here it is defined as $\dot{\nabla} = K_H^{-1} \nabla$.

Corollary 4.2 For any $F \in L^2(W \to \mathbf{R}; \mu_H)$,

$$F = \mathbf{E}[F] + \int_0^1 \mathbf{E}\left[\dot{\nabla}_s F \mid \mathcal{F}_s\right] \delta B(s).$$

Proof With the notations at hand, Theorem 4.14 implies that

$$\mathbf{E}\left[\Lambda_1^h \mid \mathcal{F}_t\right] = \exp\left(\delta_H(\pi_t^H h) - \frac{1}{2}\|\pi_t^H h\|_{\mathcal{H}_H}^2\right)$$

$$= \exp\left(\int_0^t h(s) \, \delta B(s) - \frac{1}{2}\int_0^t \dot{h}^2(s) \, ds\right).$$

This means that we have the usual relation

$$\Lambda_t^h = 1 + \int_0^t \Lambda_s \dot{h}(s)\,\delta B(s) = \mathbf{E}\left[\Lambda_1^h\right] + \int_0^1 \mathbf{E}\left[\dot{\nabla}_s \Lambda_1^h \mid \mathcal{F}_s\right]\,\delta B(s).$$

By density of the Doléans exponentials, we obtain the result.   □

Should we want to obfuscate everything, we could write

$$F = \mathbf{E}[F] + \delta_H\left(K_H\left(\mathbf{E}\left[(K_H^{-1}\nabla).F \mid \mathcal{F}.\right]\right)\right).$$

4.5 Itô Formula

Definition 4.9 Consider the operator $\mathcal{K}$ defined by $\mathcal{K} = I_{0^+}^{-1} \circ K_H$. For $H > 1/2$, it is a continuous map from $L^p([0,1] \to \mathbf{R};\ \ell)$ into $I_{H-1/2,p}$, for any $p \geq 1$. Let $\mathcal{K}_t^*$ be its adjoint in $L^p([0,t] \to \mathbf{R};\ \ell)$, i.e., for any $f \in L^p([0,1] \to \mathbf{R};\ \ell)$, any g sufficiently regular,

$$\int_0^t \mathcal{K}f(s)\,g(s)\,\mathrm{d}s = \int_0^t f(s)\,\mathcal{K}_t^* g(s)\,\mathrm{d}s.$$

The map $\mathcal{K}_t^*$ is continuous from $(I_{H-1/2,p})^*$ into $L^q([0,t] \to \mathbf{R};\ \ell)$, where $q = p/(p-1)$.

Scheme of Proof
Before going into the details of the proof of the Itô formula, we explain how it works. The basic idea is to compute

$$\lim_{\varepsilon \to 0} \varepsilon^{-1}\left(f\big(B_H(t+\varepsilon)\big) - f\big(B_H(t)\big)\right)$$

and use the fundamental theorem of calculus. As the sample-paths of B_H are nowhere differentiable, we cannot expect to use the classical chain rule formula. The idea is to work with a weak formulation, i.e., for a sufficiently rich class of nice functionals ψ, consider

$$\lim_{\varepsilon \to 0} \varepsilon^{-1}\mathbf{E}\left[\left(f\big(B_H(t+\varepsilon)\big) - f\big(B_H(t)\big)\right)\psi\right],$$

make heavy use of integration by parts until we only have classical integrals with respect to the Lebesgue measure, then take the limit, and undo the

(continued)

integration by parts to obtain a valid formula of the kind

$$\mathbf{E}\left[f(B_H(t))\psi\right] = \mathbf{E}\left[(\text{something which depends on } f' \text{ and } f'') \times \psi\right].$$

The price to pay for using such a weak approach is that the identity

$$f(B_H(t))\psi = \text{something which depends on } f' \text{ and } f'' \times \psi \qquad (4.24)$$

holds almost-surely on a set that depends on t; hence, we must take care of the continuity of all the terms of the right-hand side to construct a probability 1 set on which (4.24) holds for any t. This is the rôle of Theorem 4.16.

Along the way, to simplify some technicalities, it is well inspired to symmetrize f; hence, the introduction of the not so natural function g in (4.28).

Remark 4.6 A similar proof can be done even for $H < 1/2$ but to the price of much higher technicalities. First, $\mathcal{K}$ is no longer an integral operator but rather a fractional derivative so that the convergence of the different terms requires more stringent hypothesis on f and is harder to show. Furthermore, we must push the Taylor expansion up to n such that $2Hn > 1$ for the residual term to vanish. This means that we have terms involving increments of B_H up to the power $[1/2H] - 1$, which are handled by the same number of integrations by parts to obtain integrals with respect to Lebesgue measure (as we would differentiate a polynomial function as many times as it is necessary to obtain a constant function).

Theorem 4.15 *Assume $H > 1/2$. For $f \in C_b^2$,*

$$f(B_H(t)) = f(0) + \int_0^t \mathcal{K}_t^*(f' \circ B_H)(s)\, \delta B(s) + H\, V_H \int_0^t f''(B_H(s))s^{2H-1}\, ds.$$

Proof We begin by the symmetrization trick.

Symmetrization
Introduce the function g as

$$g(x) = f\left(\frac{a+b}{2} + x\right) - f\left(\frac{a+b}{2} - x\right). \qquad (4.25)$$

(continued)

This function is even and satisfies

$$g^{(2j+1)}(0) = 2f^{(2j+1)}(\frac{a+b}{2}) \text{ and } g(\frac{b-a}{2}) = f(b) - f(a).$$

Apply the Taylor formula to g between the points 0 and $(b-a)/2$ to get

$$f(b) - f(a) = \sum_{j=0}^{n} \frac{2^{-2j}}{(2j+1)!} (b-a)^{2j+1} f^{(2j+1)}(\frac{a+b}{2})$$

$$+ \frac{(b-a)^{2(n+1)}}{2} \int_{0}^{1} \lambda^{2n+1} g^{(2(n+1))}(\lambda a + (1-\lambda)b) \, d\lambda.$$

For any $\psi \in \mathcal{E}$ of the form $\psi = \exp(\delta_H h - \frac{1}{2}\|h\|_{\mathcal{H}_H}^2)$ with $h \in C_b^1 \subset \mathcal{H}_H$. Note that ψ satisfies $\nabla \psi = \psi h \in L^2(W \to \mathcal{H}_H; \mu_H)$. Since C_b^1 is dense into $\mathcal{H}_H$, these functionals are dense in $L^2(W \to \mathbf{R}; \mu_H)$. We then have

$$\mathbf{E}\left[\left(f(B_H(t+\varepsilon)) - f(B_H(t))\right)\psi\right]$$

$$= \mathbf{E}\left[\left(B_H(t+\varepsilon) - B_H(t)\right) f'\left(\frac{B_H(t) + B_H(t+\varepsilon)}{2}\right)\psi\right]$$

$$+\frac{1}{2}\mathbf{E}\left[\left(B_H(t+\varepsilon) - B_H(t)\right)^2 \int_0^1 r\, g^{(2)}(r B_H(t) + (1-r)B_H(t+\varepsilon))\, dr\, \psi\right]$$

$$= A_0 + \frac{1}{2}A_1. \qquad (4.26)$$

The term A_1 is the simplest to handle. If $H > 1/2$, $\varepsilon^{-1}A_1$ does vanish. Actually, recall that $B_H(t+\varepsilon) - B_H(t)$ is a centered Gaussian random variable of variance proportional to ε^{2H}; hence,

$$\varepsilon^{-1}|A_1| \le c\, \mathbf{E}\left[|B_H(t+\varepsilon) - B_H(t)|^2\right] \|f^{(2)}\|_{L^\infty}$$

$$\le c\, \varepsilon^{2H-1} \|f^{(2)}\|_{L^\infty} \xrightarrow{\varepsilon \to 0} 0,$$

since $2H - 1 > 0$.

Integration by Parts

For A_0, we have

$$A_0 = \mathbf{E}\left[\left(B_H(t+\varepsilon) - B_H(t)\right) f'\left(\frac{B_H(t) + B_H(t+\varepsilon)}{2}\right)\psi\right]$$

$$= \mathbf{E}\left[\int_0^1 \left(K_H(t+\varepsilon, s) - K_H(t, s)\right)\delta B(s)\, f'\left(\frac{B_H(t) + B_H(t+\varepsilon)}{2}\right)\psi\right]$$

$$= \mathbf{E}\left[\int_0^1 \left(K_H(t+\varepsilon, s) - K_H(t, s)\right)\dot{\nabla}_s\left(f'\left(\frac{B_H(t) + B_H(t+\varepsilon)}{2}\right)\psi\right)ds\right].$$

Since $\dot{\nabla}$ is a true derivation operator,

$$\dot{\nabla}_s\left(f'\left(\frac{B_H(t) + B_H(t+\varepsilon)}{2}\right)\psi\right) = f'\left(\frac{B_H(t) + B_H(t+\varepsilon)}{2}\right)\dot{\nabla}_s\psi$$

$$+ f''\left(\frac{B_H(t) + B_H(t+\varepsilon)}{2}\right)\left(K_H(t+\varepsilon, s) + K_H(t, s)\right).$$

Now, we only have standard integrals so that we can proceed in a classical way:

$$A_0 = \mathbf{E}\left[f'\left(\frac{B_H(t) + B_H(t+\varepsilon)}{2}\right)\int_0^1 \left(K_H(t+\varepsilon, s) - K_H(t, s)\right)\dot{\nabla}_s\psi\, ds\right]$$

$$+ \mathbf{E}\left[\psi f''\left(\frac{B_H(t) + B_H(t+\varepsilon)}{2}\right)\right.$$

$$\left.\times \int_0^1 \left(K_H(t+\varepsilon, s) - K_H(t, s)\right)\left(K_H(t+\varepsilon, s) + K_H(t, s)\right)ds\right]$$

$$= B_1 + B_2.$$

By the very definition of $\dot{\nabla}$,

$$\frac{1}{\varepsilon}\int_0^1 \left(K_H(t+\varepsilon, s) - K_H(t, s)\right)\dot{\nabla}_s\psi\, ds = \frac{1}{\varepsilon}\left(\nabla\psi(t+\varepsilon) - \nabla\psi(t)\right)$$

$$\xrightarrow{\varepsilon\to 0} \frac{d}{dt}\nabla\psi(t) = I_{0+}^{-1}\circ K_H(\dot{\nabla}\psi)(t) = \mathcal{K}(\dot{\nabla}\psi)(t).$$

Moreover, since $\nabla \psi$ belongs to $L^2(W; I_{H+1/2,2})$,

$$\mathbf{E}\left[|\nabla \psi(t+\varepsilon) - \nabla \psi(t)|^2\right] \leq c \,\|\mathcal{K}\dot{\nabla}\psi\|_{L^2(W; I_{H-1/2,2})} \,|\varepsilon|.$$

Hence,

$$\varepsilon^{-1} B_1 \xrightarrow{\varepsilon \to 0} \mathbf{E}\left[f'(B_H(t)) \,\mathcal{K}\dot{\nabla}\psi(t)\right].$$

Thanks to the symmetrization, we only have simple calculations to do for B_2:

$$B_2 = \mathbf{E}\left[\psi \, f''\left(\frac{B_H(t) + B_H(t+\varepsilon)}{2}\right)\left(R_H(t+\varepsilon, t+\varepsilon) - R_H(t, t)\right)\right]$$

and that

$$\varepsilon^{-1}\left(R_H(t+\varepsilon, t+\varepsilon) - R_H(t, t)\right) = V_H \frac{(t+\varepsilon)^{2H} - t^{2H}}{\varepsilon} \xrightarrow{\varepsilon \to 0} 2H \, V_H \, t^{2H-1}.$$

The dominated convergence theorem then yields

$$\varepsilon^{-1} B_2 \xrightarrow{\varepsilon \to 0} H \, V_H \mathbf{E}\left[\psi f''(B_H(t)) \, t^{2H-1}\right].$$

We have proved so far that

$$\frac{\mathrm{d}}{\mathrm{d}t}\mathbf{E}\left[\psi \, f(B_H(t))\right]$$

$$= \mathbf{E}\left[f'(B_H(t)) \,\mathcal{K}\dot{\nabla}\psi(t)\right] + H \, V_H \mathbf{E}\left[\psi \, f''(B_H(t)) \, t^{2H-1}\right]. \qquad (4.27)$$

It is straightforward that the right-hand side of (4.27) is continuous as a function of t on any interval $[0, T]$. Hence, we can integrate the previous relation, and we get

$$\mathbf{E}\left[\psi \, f(B_H(t))\right] - \mathbf{E}\left[\psi \, f(B_H(0))\right] = \mathbf{E}\left[\int_0^t f'(B_H(s)) \,\mathcal{K}\dot{\nabla}\psi(s) \,\mathrm{d}s\right]$$

$$+ H \, V_H \, \mathbf{E}\left[\psi \int_0^t f''(B_H(s)) \, s^{2H-1} \,\mathrm{d}s\right].$$

Remark now that

$$\mathbf{E}\left[\int_0^t f'\big(B_H(s)\big)\,\mathcal{K}\dot{\nabla}\psi(s)\,\mathrm{d}s\right] = \mathbf{E}\left[\int_0^1 f'\big(B_H(s)\big)\mathbf{1}_{[0,t]}(s)\,\mathcal{K}\dot{\nabla}\psi(s)\,\mathrm{d}s\right]$$

$$= \mathbf{E}\left[\int_0^1 \mathcal{K}_1^*\big(f' \circ B_H\,\mathbf{1}_{[0,t]}\big)\,\dot{\nabla}_s\psi\,\mathrm{d}s\right] = \mathbf{E}\left[\psi\int_0^1 \mathcal{K}_1^*\big(f' \circ B_H\,\mathbf{1}_{[0,t]}\big)(s)\,\delta B(s)\right].$$

Note that

$$\mathcal{K}_1^*(f'\mathbf{1}_{[0,t]})(s) = \frac{\mathrm{d}}{\mathrm{d}s}\int_s^1 K(r,s)f'(r)\mathbf{1}_{[0,t]}(r)\,\mathrm{d}r = 0 \text{ if } s > t.$$

This means that

$$\pi_t^H\left(\mathcal{K}_t^*(f'\mathbf{1}_{[0,t]})\right) = \mathcal{K}_t^*(f'\mathbf{1}_{[0,t]})$$

and by the definition (4.23),

$$\int_0^1 \mathcal{K}_t^*\big(f \circ B_H\,\mathbf{1}_{[0,t]}\big)(s)\,\delta B(s) = \int_0^t \mathcal{K}_t^*\big(f \circ B_H\big)(s)\,\delta B(s).$$

Consequently, we have

$$\mathbf{E}\left[\psi\,f\big(B_H(t)\big)\right] - \mathbf{E}\left[\psi\,f\big(B_H(0)\big)\right] = \mathbf{E}\left[\psi\int_0^t \mathcal{K}_t^*(f \circ B_H)(s)\delta B(s)\right]$$

$$+ H\,V_H\,\mathbf{E}\left[\psi\int_0^t f''\big(B_H(s)\big)\,s^{2H-1}\,\mathrm{d}s\right].$$

Since the functionals ψ we considered form a dense subset in $L^2(\mathrm{W} \to \mathbf{R};\,\mu_H)$, we have

$$f\big(B_H(t)\big) - f\big(B_H(0)\big) = \int_0^t \mathcal{K}_t^*(f \circ B_H)(s)\delta B(s)$$

$$+ H\,V_H\int_0^t f''\big(B_H(s)\big)\,s^{2H-1}\,\mathrm{d}s, \quad \mathrm{d}t \otimes \mu_H\text{-a.s.} \qquad (4.28)$$

Admit for a while that

$$t \longrightarrow \int_0^t \mathcal{K}_t^*(f' \circ B_H)(s)\delta B(s)$$

has almost-surely continuous sample-paths. It is clear that the other terms of (4.28) have also continuous trajectories. Let A be the negligeable set of $W \times [0,1]$

where (4.28) does not hold. According to the Fubini theorem, for any $t \in [0, 1]$, the set

$$A_t = \{\omega \in W, \ (\omega, t) \in\} \in A\}$$

is negligeable and so does $A_{\mathbf{Q}} = \cup_{t \in [0,1] \cap \mathbf{Q}} A_t$. For any $t \in \mathbf{Q} \cap [0, 1]$, Eq. (4.28) holds on $A_{\mathbf{Q}}^c$, i.e., holds μ_H-almost-surely. By continuity, this is still true for any $t \in [0, 1]$.   □

Theorem 4.16 *For any $H \in [1/2, 1)$. Let u belong to $\mathbb{D}_{p,1}^H(L^p)$ with $Hp > 1$. The process*

$$U(t) = \int_0^t \mathcal{K}_t^* u(s) \delta B(s), \ t \in [0, 1]$$

admits a modification with $(H - 1/p)$-Hölder continuous paths, and we have the maximal inequality:

$$\mathbf{E} \left[\sup_{r \neq t \in [0,1]^2} \frac{\left| \int_0^1 \left(\mathcal{K}_t^* u(s) - \mathcal{K}_r^* u(s) \right) \delta B(s) \right|^p}{|t - r|^{pH}} \right]^{1/p} \leq c \|\mathcal{K}_1^*\|_{H,2} \|u\|_{\mathbb{D}_{p,1}^H}.$$

Proof For $g \in C^\infty$ and ψ a cylindric real-valued functional,

$$\mathbf{E} \left[\int_0^1 \int_0^t \mathcal{K}_t^* u(s) \delta B(s) \ g(t) \, dt \ \psi \right] = \mathbf{E} \left[\iint_{[0,1]^2} \mathcal{K}_1^* (u \mathbf{1}_{[0,t]})(r) g(t) \dot{\nabla}_r \psi \, dt \, dr \right]$$

$$= \mathbf{E} \left[\int_0^1 \mathcal{K}_1^* (u I_{1-}^1 g)(r) \dot{\nabla}_r \psi \, dr \right] = \mathbf{E} \left[\delta(\mathcal{K}_1^* (u . I_{1-}^1 g) \psi) \right].$$

Thus,

$$\int_0^1 \int_0^t \mathcal{K}_t^* u(s) \, \delta B(s) \ g(t) \, dt = \int_0^1 \mathcal{K}_1^* (u . I_{1-}^1 g)(s) \, \delta B(s) \ \mu_H - \text{a.s.} \tag{4.29}$$

Since $H > 1/2$, it is clear that $\mathcal{K}$ is continuous from $L^2([0, 1] \to \mathbf{R}; \ell)$ into $I_{H-1/2,2}$ and thus that $\mathcal{K}_1^*$ is continuous from $I_{H-1/2,2}^*$ in $L^2([0, 1] \to \mathbf{R}; \ell)$. Since $I_{H-1/2,2}$ is continuously embedded in $L^{(1-H)^{-1}}([0, 1] \to \mathbf{R}; \ell)$, it follows that $L^{1/H}([0, 1] \to \mathbf{R}; \ell) = (L^{(1-H)^{-1}}([0, 1] \to \mathbf{R}; \ell))^*$ is continuously embedded in $I_{1/2-H,2}$. Since u belongs to $\mathbb{D}_{p,1}^H(L^p)$, the generalized Hölder inequality implies that

$$\|u I_{1-}^1 g\|_{L^{1/H}} \leq \|u\|_{L^p} \|I_{1-}^1 g\|_{L^{(H-1/p)^{-1}}}.$$

It follows that U belongs to $L^p\big(\mathbf{W} \to I^+_{1,(1-H+1/p)^{-1}};\ \mu_H\big)$ with

$$\|U\|_{L^p\big(\mathbf{W} \to I^+_{1,(1-H+1/p)^{-1}};\ \mu_H\big)} \le c\|\mathcal{K}^*_1\|_{H,2}\|u\|_{\mathbb{D}^H_{p,1}}.$$

The proof is completed remarking that $1 - 1/(1 - H + 1/p)^{-1} = H - 1/p$ so that $I^+_{1,(1-H+1/p)^{-1}}$ is embedded in $\mathrm{Hol}(H - 1/p)$.   □

Deterministic Fractional Calculus

We now consider some basic aspects of the deterministic fractional calculus—the main reference for this subject is [8].

Definition 4.10 Let $f \in L^1\big([a, b] \to \mathbf{R};\ \ell\big)$, the integrals

$$(I^\alpha_{a+} f)(x) = \frac{1}{\Gamma(\alpha)} \int_a^x f(t)(x - t)^{\alpha-1}\, dt\ ,\ x \ge a,$$

$$(I^\alpha_{b-} f)(x) = \frac{1}{\Gamma(\alpha)} \int_x^b f(t)(x - t)^{\alpha-1}\, dt\ ,\ x \le b,$$

where $\alpha > 0$, are, respectively, called right and left fractional integrals of the order α.

For any $\alpha \ge 0$, any $f \in L^p\big([0, 1] \to \mathbf{R};\ \ell\big)$ and $g \in L^q\big([0, 1] \to \mathbf{R};\ \ell\big)$ where $p^{-1} + q^{-1} \le \alpha$, we have

$$\int_0^t f(s)(I^\alpha_{0+} g)(s)\, ds = \int_0^t (I^\alpha_{t-} f)(s)g(s)\, ds. \tag{4.30}$$

Moreover, the family of fractional integrals constitute a semi-group of transformations: For any $\alpha, \beta > 0$,

$$I^\alpha_{0+} \circ I^\beta_{0+} = I^{\alpha+\beta}_{0+}. \tag{4.31}$$

Definition 4.11 For f given in the interval $[a, b]$, each of the expressions

$$(\mathcal{D}^\alpha_{a+} f)(x) = \left(\frac{d}{dx}\right)^{[\alpha]+1} I^{1-\{\alpha\}}_{a+} f(x),$$

$$(\mathcal{D}^\alpha_{b-} f)(x) = \left(-\frac{d}{dx}\right)^{[\alpha]+1} I^{1-\{\alpha\}}_{b-} f(x)$$

are, respectively, called the right and left fractional derivatives (provided they exist), where $[\alpha]$ denotes the integer part of α and $\{\alpha\} = \alpha - [\alpha]$.

Theorem 4.17 *We have the following embeddings and continuity results:*

1. *If $0 < \gamma < 1$, $1 < p < 1/\gamma$, then I_{0+}^{γ} is a bounded operator from $L^p([0, 1] \to$ $\mathbf{R}; \ell)$ into $L^q([0, 1] \to \mathbf{R}; \ell)$ with $q = p(1 - \gamma p)^{-1}$.*
2. *For any $0 < \gamma < 1$ and any $p \geq 1$, $I_{\gamma,p}^{+}$ is continuously embedded in $\mathrm{Hol}(\gamma - 1/p)$ provided that $\gamma - 1/p > 0$.*
3. *For any $0 < \gamma < \beta < 1$, $\mathrm{Hol}(\beta)$ is compactly embedded in $I_{\gamma,\infty}$.*

4.6   Problems

4.1 (About Causality) Let V be a causal operator from $L^2([0, 1] \to \mathbf{R}; \ell)$ into itself. Let

$$V_t = \dot{\pi}_t \circ V \circ \dot{\pi}_t \; : \; L^2([0, 1] \to \mathbf{R}; \ell) \longrightarrow L^2([0, t] \to \mathbf{R}; \ell)$$

$$f \longmapsto V(f \mathbf{1}_{[0,t]}) \mathbf{1}_{[0,t]}.$$

Let V_t^* be the adjoint of V_t.

1. Show that V_t^* is continuous from $L^2([0, t] \to \mathbf{R}; \ell)$ into $L^2([0, 1] \to \mathbf{R}; \ell)$. (We here identify $L^2([0, 1] \to \mathbf{R}; \ell)$ with its dual)

Consider the situation where

$$Vf(r) = \int_0^t V(r, s) f(s) \, ds$$

with $V(r, s) = 0$ whenever $s > r$. Note that this is the case of $\mathcal{K}_H$ for $H > 1/2$.

1. Show that

$$V_t^* f = V_1^*(\dot{\pi}_t f).$$

2. Derive the same identity using solely the causality of V.

$V = \mathcal{K}_H$ for $H < 1/2$ corresponds to this last situation.

4.2 (Riemann Sums for fBm) One approach to define a stochastic integral with respect to B_H for $H > 1/2$ is to look at Riemann -like sums:

$$RS_n(U) = \sum_{i=0}^{n-1} U(i/n) \left(B_H \left(\frac{i+1}{n} \right) - B_H \left(\frac{i}{n} \right) \right)$$

Consider that $U(s) = \delta_H h\, u(s)$ where u is deterministic and continuous on $[0, 1]$ and h is C^1, hence belonging to $\mathcal{H}_H$.

1. Show that

$$\dot{\nabla}_r U(s) = u(s)\dot{h}(r)$$

where $\dot{h} = K_H^{-1}(h)$.

2. Derive

$$\left(K_{1/2}^{-1} \circ K_H \circ \dot{\nabla}\right)_r \dot{U}(s) = u(s)h'(r).$$

3. Show that

$$RS_n(U) = \int_0^1 \sum_{i=0}^{n-1} U(\frac{i}{n}) \left(K_H(\frac{i+1}{n}, r) - K_H(\frac{i}{n}, r)\right) \delta B(r)$$

$$+ \sum_{i=0}^{n-1} u(\frac{i}{n}) \left(h(\frac{i+1}{n}) - h(\frac{i}{n})\right).$$

4. Assume for the next two questions only that K_H is regular as it needs to be. Show that

$$\sum_{i=0}^{n-1} U(\frac{i}{n}) \left(K_H(\frac{i+1}{n}, r) - K_H(\frac{i}{n}, r)\right) \xrightarrow{n\to\infty} \int_0^1 U(s)\frac{d}{ds} K_H(\varepsilon_r)(s)\, ds$$

where ε_r is the Dirac measure at r.

5. Derive the following identity:

$$\int_0^1 U(s)\frac{d}{ds} K_H(\varepsilon_r)(s)\, ds = \widehat{\mathcal{K}_H}^* U(r),$$

where $\widehat{\mathcal{K}_H} = K_{1/2}^{-1} \circ K_H$.

6. Show that

$$\sum_{i=0}^{n-1} u(\frac{i}{n}) \left(h(\frac{i+1}{n}) - h(\frac{i}{n})\right) \xrightarrow{n\to\infty} \int_0^1 u(s)h'(s)\, ds = \text{trace}\left(\widehat{\mathcal{K}_H}\dot{\nabla}U\right).$$

The map $\widehat{\mathcal{K}_H} = K_{1/2}^{-1} \circ K_H$ is a continuous map from $L^2([0, 1] \to \mathbf{R}; \ell)$ into $I_{H-1/2,2}$ so that a possible definition of a stochastic integral (in the sense of Riemann integrals) could be

$$\delta_H(\widehat{\mathcal{K}_H}^* U) + \text{trace}(\widehat{\mathcal{K}_H}\dot{\nabla}U)$$

provided that U has the necessary regularity for these terms to make sense.

4.7 Notes and Comments

The paper [5] was the first to construct the Malliavin calculus for fractional Brownian motion. The activity on this subject has been frantic during the first ten years of the millennium. One question was to establish an Itô formula for the smallest possible value of H. The proof here is done for $H > 1/2$ for the sake of simplicity but can be adapted (to the price of an increased complexity) to any $H \in (0, 1/2)$ (see [4]). In the end, the Itô formula for fBm is not as fruitful as its counterpart for the ordinary Brownian motion, since it cannot be read as a stability result: the operators that appear in the right-hand side of the Itô formula are not local but more of the sort of integro-differential maps.

There exists another presentation of the Cameron–Martin space of the fBm in [7]; the similarity and difference between the two approaches are explained in [3, Chapter 10].

The last difficulty encountered with the fBm is that the divergence cannot be considered as a stochastic integral in the usual sense as it does not coincide with any limit of Riemann-like or Stratonovich-like sums. All these constructions lead to a trace term whose existence itself requires strong hypothesis on the integrand.

References

1. M.A. Arcones, On the law of the iterated logarithm for Gaussian processes. J. Theor. Probab. **8**(4), 877–903 (1995)
2. R.J. Barton, H. Vincent Poor, Signal detection in fractional Gaussian noise. IEEE Trans. Inf. Theory **34**(5), 943–959 (1988)
3. L. Decreusefond, Stochastic integration with respect to fractional Brownian motion, in *Theory and Applications of Long-Range Dependence* (Birkhäuser, Boston, 2003), pp. 203–226
4. L. Decreusefond, Stochastic calculus with respect to Volterra processes. Ann. l'Institut Henri Poincaré (B) Probab. Stat. **41**, 123–149 (2005)
5. L. Decreusefond, A.S. Üstünel, Stochastic analysis of the fractional Brownian motion. Potential Anal. **10**(2), 177–214 (1999)
6. A.F. Nikiforov, V.B. Uvarov, *Special Functions of Mathematical Physics* (Birkhäuser, Boston, 1988)
7. D. Nualart, *The Malliavin Calculus and Related Topics* (Springer, Berlin, 1995)
8. S.G. Samko, A.A. Kilbas, O.I. Marichev, *Fractional Integrals and Derivatives* (Gordon and Breach Science, Philadelphia, 1993)

Chapter 5
Poisson Space

Abstract The Poisson process on the half-line shares many properties with the Brownian motion due to the fact that it also has stationary and independent increments. As such, it has been the second process for which a Malliavin structure has been constructed. It turns out that the underlying time scale is not necessary to develop this theory. Hence, we consider Poisson point processes in (almost) any topological space.

5.1 Point Processes

Let us define what is a configuration, the basic element of our random experiments, which play the rôle of the trajectories of the Brownian motion.

Definition 5.1 Let E be a metrizable, separable, and complete space, i.e., a Polish space (actually we could be more General, but it is of no use here). A configuration is a locally finite set (i.e., there is a finite number of points in any bounded set) of points of a set E. We denote $\mathfrak{N}_E$ the set of configurations of E. A generic element of $\mathfrak{N}_E$ is then a sequence $\phi = (x_n, \ n \geq 1)$ of elements of E.

> **Set or Measure?**
> It is often convenient to see configurations as atomic measures: We can view the set $\phi = (x_n, \ n = 1, \cdots, M)$ (where $M \in \mathbf{N} \cup \{+\infty\}$) as the measure
>
> $$\phi = \sum_{n=1}^{M} \varepsilon_{x_n}$$
>
> where ε_a is the Dirac mass at point a. We abuse the notation and keep the same letter ϕ for both descriptions. In order to keep in mind that there is no privileged order in the enumeration of the elements of ϕ, we prefer to write

(continued)

© The Author(s), under exclusive license to Springer Nature Switzerland AG 2022
L. Decreusefond, *Selected Topics in Malliavin Calculus*,
Bocconi & Springer Series 10, https://doi.org/10.1007/978-3-031-01311-9_5

$$\sum_{x \in \phi} \varepsilon_x \text{ instead of } \sum_{n=1}^{M} \varepsilon_{x_n}.$$

When we want to count the number of points of ϕ that fall in a subset A, we can alternatively write

$$\phi(A) = \text{card}\{x \in \phi,\ x \in A\} = \int_A d\phi(x).$$

For $A \subset E$, we denote by ϕ_A the restriction of ϕ to A:

$$\phi_A = \{x \in \phi,\ x \in A\} = \sum_{x \in \phi} \mathbf{1}_A(x)\, \varepsilon_x.$$

To make $\mathfrak{N}_E$ a topological space, we furnish it with the topology induced by the semi-norms

$$p_f(\phi) := \left| \int_E f \, d\phi \right| = \left| \sum_{x \in \phi} f(x) \right|$$

for $f \in C_K(E \to \mathbf{R})$, the set of continuous functions with compact support from E to $\mathbf{R}$. This means that

$$\phi_n \overset{\text{vaguely}}{\longrightarrow} \phi \iff p_f(\phi - \phi_n) \xrightarrow{n \to \infty} 0, \ \forall f \in C_K(E \to \mathbf{R}).$$

Then, $\mathfrak{N}_E$ is in turn a metrizable, separable, and complete space.

Remark 5.1 The locally finite hypothesis entails that a configuration is a finite or denumerable set of points of E. However, a set like $\{1/n,\ n \geq 1\}$ is not a configuration in $E = [0, 1]$ since 0 is an accumulation point.

Remark 5.2 The vague convergence of ϕ_n toward ϕ means that each atom of ϕ is the limit of a sequence of atoms of ϕ_n. However, since the test functions that define the semi-norms have compact support, there is no uniformity in this convergence. For instance, the sequence $(\varepsilon_n,\ n \geq 1)$ converges vaguely to the null measure.

Definition 5.2 A point process N is an $\mathfrak{N}_E$-valued random variable.

According to the general theory of points processes, the rôle of the characteristic function is played by the Laplace transform Φ_N.

Definition 5.3 For N a point process on a Polish space E, its Laplace transform is defined by

$$\Phi_N : f \in C_K(E \rightarrow \mathbf{R}) \longmapsto \mathbf{E}\left[\exp\left(-\int_E f \, dN\right)\right].$$

Theorem 5.1 *Let N and N' be two point processes on E. Then, they have the same distribution if and only if $\Phi_N = \Phi_{N'}$.*

Example (Bernoulli Point Process) The Bernoulli point process is a process based on a finite set $E = \{x_1, \cdots, x_m\}$. We introduce $X_1, \cdots, X_m$ some random independent variables of Bernoulli distribution with parameter p. The Bernoulli point process is then defined by

$$N = \sum_{i=1}^n X_i \, \varepsilon_{x_i}.$$

Example (Binomial Process) The number of points is fixed to m, and a probability measure $\tilde{\sigma}$ on E is given. The m atoms are independently drawn randomly according to $\tilde{\sigma}$. It is straightforward that

$$\mathbf{P}\big(N(A) = k\big) = \binom{m}{k} \tilde{\sigma}(A)^k \big(1 - \tilde{\sigma}(A)\big)^{m-k}$$

and for $A_1, \cdots, A_n$, a partition of E and $(k_1, \cdots, k_n)$ such that $\sum_{i=1}^n k_i = m$,

$$\mathbf{P}\big(N(A_1) = k_1, \cdots, N(A_n) = k_n\big) = \frac{m!}{k_1! \ldots k_n!} \, \tilde{\sigma}(A_1)^{k_1} \ldots \tilde{\sigma}(A_n)^{k_n}. \qquad (5.1)$$

Theorem 5.2 *The Laplace transform of the binomial process is given by*

$$\mathbf{E}\left[\exp\left(-\int_E f \, dN\right)\right] = \exp\left(-m \int_E f(x) \, d\tilde{\sigma}(x)\right).$$

Proof Denote by $(X_1, \cdots, X_n)$ the locations of the points of N. By independence, we have

$$\mathbf{E}\left[\exp\left(-\int_E f \, dN\right)\right] = \mathbf{E}\left[\exp\left(-\sum_{i=1}^m f(X_i)\right)\right]$$

$$= \prod_{i=1}^m \exp\left(-\int_E f(x) \, d\tilde{\sigma}(x)\right).$$

Hence, the result. $\qquad\qquad\square$

5.2   Poisson Point Process

The point process, mathematically the richest, is the spatial Poisson process that generalizes the Poisson process on the real line (see Sect. 5.5 for some very quick refresher on the Poisson process on $\mathbf{R}^+$). It is defined as a binomial point process with a random number of points M, independent of the locations. The distribution of the number of points is chosen to be Poisson for the process to have nice properties. This amounts to say that we consider the probability space

$$\Omega = \mathbf{N} \times E^{\mathbf{N}}$$

equipped with the measure

$$\left(\sum_{m=0}^{\infty} e^{-a} \frac{a^n}{n!} \, \varepsilon_n \right) \otimes \tilde{\sigma}^{\otimes \mathbf{N}}. \tag{5.2}$$

The process N is defined as the map

$$N \; : \; \Omega = \mathbf{N} \times E^{\mathbf{N}} \longrightarrow \mathfrak{N}_E$$

$$\omega = (m, x_1, x_2, \cdots) \longmapsto \sum_{k=1}^{m} \varepsilon_{x_k}$$

with the convention that $\sum_{k=1}^{0} \cdots = \emptyset$. It is then straightforward that

$$\mathbf{E}\left[\exp\left(-\int_E f \, dN \right) \right] = \sum_{m=0}^{\infty} \exp\left(-m \int_E f(x) d\sigma(x) \right) \mathbf{P}(M = m)$$

$$= \exp\left(-\int_E \left(1 - e^{-f(x)}\right) a \, d\tilde{\sigma}(x) \right).$$

This leads to the following definition.

Definition 5.4 Let σ be a finite measure on a Polish space E. The Poisson process with intensity σ, denoted by N, is defined by its Laplace transform: for any function $f \in C_K(E \to \mathbf{R})$,

$$\Phi_N(f) = \exp\left(-\int_E \left(1 - e^{-f(x)}\right) d\sigma(x) \right). \tag{5.3}$$

We denote by π^{σ} the Poisson measure of intensity σ that is the law of the Poisson process of intensity σ.

Remark 5.3 To construct the Poisson measure of intensity σ, set $a = \sigma(E)$ and $\tilde{\sigma} = a^{-1}\sigma$ in (5.2).

5.3   Finite Poisson Point Process

The general definition of a Poisson point process does not need that its intensity is a finite measure. For the sake of simplicity, we here assume that

$$\sigma(E) < \infty.$$

We also assume that σ is diffuse, i.e., $\sigma(\{x\}) = 0$ for any $x \in E$.

As usual with Laplace transforms, by derivation, we obtain expression of moments. The subtlety lies in the *diagonal* terms: For an integral with respect to the Lebesgue measure, the diagonal does not weigh

$$\left(\int_{\mathbf{R}} f(s) ds \right)^2 = \iint_{\mathbf{R} \times \mathbf{R}} f(s) f(t) \, ds dt = \iint_{\mathbf{R} \times \mathbf{R} \setminus \Delta} f(s) f(t) \, ds dt$$

where $\Delta = \{(x, y) \in \mathbf{R}^2, \ x = y\}$. When we have integrals with respect to atomic measures, we must take care of the diagonal terms:

$$\left(\sum_{x \in \phi} f(x) \right)^2 = \sum_{x \in \phi, y \in \phi} f(x) f(y)$$

$$= \sum_{x \in \phi, y \in \phi, x \neq y} f(x) f(y) + \sum_{x \in \phi} f(x)^2.$$

We thus introduce the notation

$$\phi_{\neq}^{(2)} = \{(x, y) \in \phi \times \phi, \ x \neq y\}.$$

Theorem 5.3 (Campbell Formula) *Let $f \in L^1(E \to \mathbf{R}; \sigma)$,*

$$\mathbf{E}\left[\int_E f dN \right] = \int_E f d\sigma \tag{5.4}$$

and if $f \in L^2(E \times E \to \mathbf{R}; \sigma \otimes \sigma)$, then

$$\mathbf{E}\left[\sum_{x, y \in N_{\neq}^{(2)}} f(x, y) \right] = \iint_{E \times E} f(x, y) d\sigma(x) d\sigma(y). \tag{5.5}$$

Proof By the very definition of N, for any θ, we have

$$\mathbf{E}\left[\exp\left(-\theta \int_E f dN \right) \right] = \exp\left(-\int_E \left(1 - e^{-\theta f(x)} \right) d\sigma(x) \right).$$

On the one hand,

$$\frac{d}{d\theta} \mathbf{E}\left[\exp\left(-\theta \int_E f dN\right)\right] = -\mathbf{E}\left[\int_E f dN \, \exp\left(-\theta \int_E f dN\right)\right],$$

and on the other hand,

$$\frac{d}{d\theta} \exp\left(-\int_E \left(1 - e^{-\theta f(x)}\right) d\sigma(x)\right)$$

$$= -\int_E f(x) e^{-\theta f(x)} d\sigma(x) \, \exp\left(-\int_E \left(1 - e^{-\theta f(x)}\right) d\sigma(x)\right).$$

Take $\theta = 0$ to obtain (5.4).

Similarly,

$$\frac{d^2}{d\theta^2} \mathbf{E}\left[\exp\left(-\theta \int_E f dN\right)\right] = \mathbf{E}\left[\left(\int_E f dN\right)^2 \exp\left(-\theta \int_E f dN\right)\right]$$

and

$$\frac{d^2}{d\theta^2} \exp\left(-\int_E \left(1 - e^{-\theta f(x)}\right) d\sigma(x)\right)$$

$$= \left[\left(\int_E f(s) e^{-\theta f(x)} d\sigma(x)\right)^2 + \int_E f(x)^2 d\sigma(x)\right]$$

$$\times \exp\left(-\int_E \left(1 - e^{-\theta f(x)}\right) d\sigma(x)\right).$$

For $\theta = 0$, we obtain

$$\mathbf{E}\left[\left(\int_E f dN\right)^2\right] = \left(\int_E f(s) e^{-\theta f(x)} d\sigma(x)\right)^2 + \int_E f(x)^2 d\sigma(x).$$

By the definition of the stochastic integral with respect to N:

$$\left(\int_E f dN\right)^2 = \sum_{x \in N, y \in N} f(x) f(y) = \sum_{x, y \in N_{\neq}^{(2)}} f(x) f(y) + \sum_{x \in N} f(x)^2.$$

From the first part of the proof, we know that

$$\mathbf{E}\left[\sum_{x \in N} f(x)^2\right] = \int_E f(x)^2 d\sigma(x).$$

Hence,

$$\mathbf{E}\left[\sum_{x,y\in N^{\neq}} f(x)f(y)\right] = \int_{E^2} f(x)f(y)d\sigma(x)d\sigma(y).$$

Then, (5.5) follows by polarization and density of simple tensor products in $L^2(E \times E \to \mathbf{R}; \sigma \otimes \sigma)$ □

An alternative definition of the Poisson process is as follows:

Theorem 5.4 *A point process N is a Poisson process with intensity σ if and only if:*

(i) For every set $K \subset E$, $N(K)$ follows a Poisson distribution with parameter $\sigma(K)$.

(ii) For K_1 and K_2, two disjoint subsets of $(E, \mathcal{B}(E))$, the random variables $N(K_1)$ and $N(K_2)$ are independent.

Proof **Step 1** Consider $f = \theta_1 \mathbf{1}_{K_1} + \theta_2 \mathbf{1}_{K_2}$. If $K_1 \cap K_2 = \emptyset$, then

$$e^{-f(x)} = e^{-\theta_1}\mathbf{1}_{K_1}(x) + e^{-\theta_2}\mathbf{1}_{K_2}(x) + \mathbf{1}_{(K_1 \cup K_2)^c}(x). \tag{5.6}$$

Then, according to (5.3),

$$\mathbf{E}\left[e^{-\theta_1 N(K_1)}e^{-\theta_2 N(K_2)}\right]$$
$$= \exp\left(-\left(\sigma(E) - e^{-\theta_1}\sigma(K_1) - e^{-\theta_2}\sigma(K_2) - \sigma((K_1 \cup K_2)^c)\right)\right)$$
$$= \prod_{i=1,2} \exp\left(\sigma(K_i) - e^{-\theta_i}\sigma(K_i)\right).$$

We recognize the product of the Laplace transforms of two independent Poisson random variables of the respective parameters $\sigma(K_1)$ and $\sigma(K_2)$.

Step 2 In the converse direction, assume that the properties 5.4 and 5.4 hold true. Consider f a step function:

$$f(x) = \sum_{i=1}^{n} \theta_i \mathbf{1}_{K_i}$$

where $(K_i, 1 \leq i \leq n)$ are the measurable sets of E, two by two disjoint. By independence,

$$\mathbf{E}\left[\exp\left(-\int_E f dN\right)\right] = \prod_{j=1}^{n} \mathbf{E}\left[\exp\left(-\theta_i N(K_i)\right)\right].$$

Since $N(K_i)$ is a Poisson random variable of parameter $\sigma(K_i)$, we get

$$\mathbf{E}\left[\exp\left(-\int_E f\,dN\right)\right] = \prod_{j=1}^{n} \exp\left(\sigma(K_i) - e^{-\theta_i}\sigma(K_i)\right).$$

Using the trick of (5.6), we see that

$$\mathbf{E}\left[\exp\left(-\int_E f\,dN\right)\right] = \exp\left(-\int_E (1 - e^{-f})d\sigma\right) \qquad (5.7)$$

for non-negative finite-valued functions. By monotone convergence, (5.7) still holds for non-negative measurable functions; hence, N is a Poisson process.

$\square$

5.3.1 Operations on Configurations

There are a few transformations that can be made on configurations.

The superposition of ϕ_1 and ϕ_2 is the union of the two sets counting the points with multiplicity or more clearly the sum of the two measures.

For p a map from E to $[0, 1]$, the p-thinning of $\phi = \{x_1, \cdots, x_n\}$ is the random measure

$$p \circ \phi := \sum_{i=1}^{n} \mathbf{1}_{\{U_i \leq p(x_i)\}} \varepsilon_{x_i}$$

where $(U_i, i \geq 1)$ is a family of independent uniform random variables over $[0, 1]$.

If E is a cone, i.e., if we can multiply each $x \in E$ by a non-negative scalar a, then the dilation of ϕ is the configuration whose atoms are $\{ax, x \in E\}$.

It is clear from (5.3) that the following theorem holds.

Theorem 5.5 *Let N^1 and N^2 be two independent Poisson processes with the respective intensities σ^1 and σ^2, their superposition N is a Poisson process with intensity $\sigma^1 + \sigma^2$.*

Theorem 5.6 *A p-thinned Poisson process of intensity σ is a Poisson process of intensity σ_p defined by*

$$\sigma_p(A) = \int_A p(x)d\sigma(x).$$

Proof We have to prove that

$$\mathbf{E}\left[e^{-\int_E f\,d(p\circ N)}\right] = \exp\left(-\int_E (1 - e^{-f})p\,d\sigma\right). \qquad (5.8)$$

For Y a $0/1$ Bernoulli random variable of success probability p, let

$$L_Y(t) = \mathbf{E}\left[e^{sY}\right] = e^s p + (1-p) := l(s, p).$$

We denote by $(Y_x, x \in E)$ a family of the Bernoulli random variables that decides whether we keep the atom located at x. For the sake of notations, we denote temporarily $\sigma_n = \pi^\sigma(N(E) = n)$.

$$\mathbf{E}\left[e^{-\int_E f \mathrm{d}(p \circ N)}\right] = \mathbf{E}\left[e^{-\sum_{x \in N} Y_x f(x)}\right]$$

$$= 1 + \sum_{n=1}^{\infty} \mathbf{E}\left[e^{-\sum_{x \in N} Y_x f(x)} \mid N(E) = n\right] \sigma_n$$

$$= 1 + \sum_{n=1}^{\infty} \prod_{j=1}^{n} \frac{1}{\sigma(E)} \int_E \mathbf{E}\left[e^{-Y_{x_j} f(x_j)}\right] \mathrm{d}\sigma(x_j)\, \sigma_n$$

$$= 1 + \sum_{n=1}^{\infty} \prod_{j=1}^{n} \frac{1}{\sigma(E)} \int_E l(-f(x_j), p(x_j))\, \mathrm{d}\sigma(x_j)\, \sigma_n$$

$$= 1 + \sum_{n=1}^{\infty} \exp\left(\sum_{j=1}^{n} \frac{1}{\sigma(E)} \log \int_E l(f(-x_j), p(x_j))\, \mathrm{d}\sigma(x_j)\right) \sigma_n$$

$$= \mathbf{E}\left[\exp\left(\int_E \log l(-f(x), p(x)) \mathrm{d}N(x)\right)\right]$$

$$= \exp\left(-\int_E 1 - l(f(x), p(x)) \mathrm{d}\sigma(x)\right)$$

$$= \exp\left(-\int_E \left(1 - e^{-f(x)}\right) p(x) \mathrm{d}\sigma(x)\right),$$

which is (5.8). $\qquad\qquad\square$

Example (M/M/∞ Queue) The M/M/∞ queue is the queue with Poisson arrivals, independent and identically distributed from exponential distribution service times, and an infinite number of servers (without buffer). It is initially a theoretical object that is particularly simple to analyze and also a model to which we can compare other situations.

The process of interest is X that counts the number of occupied servers. It may be studied through the framework of continuous time Markov chains but with some difficulties since the coefficients of the infinitesimal generator are not bounded so that the associated semi-group is not continuous from $l^\infty(\mathbf{N})$ into itself.

Fig. 5.1 The graphical
representation of the domain
of integration

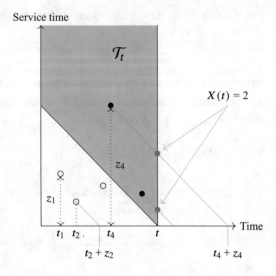

Let $(t_n, n \geq 1)$ be the arrival times and $(z_n, n \geq 1)$ the service times. This means
that the n-th customer arrives at t_n and leaves the system at time $t_n + z_n$. It is fruitful
to represent this phenomenon by the following picture, see Fig. 5.1.

This representation means that a customer who arrives at $s < t$ is still in the
system at t if and only its service duration is larger than $t - s$. This corresponds to
points in the upper trapezoid

$$\mathcal{T}_t = \left\{ (s, z) \in \mathbf{R}^+ \times \mathbf{R}^+, \, 0 \leq s \leq t, \, z \geq t - s \right\}.$$

Consider that the arrivals occur according to a Poisson process on the half-line
of intensity $\sigma = \rho \ell$ where $\rho > 0$ and that the service times follow an exponential
distribution of parameter 1. The number of points in the rectangle $[0, t] \times \mathbf{R}^+$ is the
number of arrivals before t. A customer arrived before t has its representative point
in the upper trapezoid with probability

$$\mathbf{P}(\text{service time} > t - \text{arrival time}) = \exp\left(- (t - \text{arrival time}) \right).$$

Hence, the number of points in $\mathcal{T}_t$ is the p_t-thinning of the arrival process where

$$p_t(s) = \exp\left(- (t - s) \right).$$

According to 5.6, this means that $X(t)$ follows a Poisson distribution of parameter

$$\int_0^t \exp\left(- (t - s) \right) \rho \, ds = \rho (1 - e^{-t}). \tag{5.9}$$

Definition 5.5 For X an integer valued random variable and $p \in [0, 1]$, the p-thinning of X is the random variable, and we also denote by $p \circ X$ (as there is no risk of confusion with the thinning of a configuration) defined by

$$p \circ X \stackrel{\text{dist.}}{=} \sum_{j=1}^{X} B_j$$

where $(B_j, j \geq 1)$ is a family of independent (and independent of X) Bernoulli random variables of success parameter p. By convention, $\sum_{j=1}^{0} \ldots = 0$.

A short computation shows that:

Lemma 5.1 *If X is a Poisson random variable of parameter λ, then $p \circ X$ is distributed as a Poisson distribution of parameter λp.*

Proof Compute the generating function of $p \circ X$:

$$\mathbf{E}\left[s^{p \circ X}\right] = \sum_{k=0}^{\infty} \mathbf{E}\left[\prod_{j=1}^{k} s^{B_j}\right] mathbf{P}(X = k)$$

$$= e^{-\lambda} \sum_{k=0}^{\infty} (ps + 1 - p)^k \frac{\lambda^k}{k!}$$

$$= \exp\left(-\lambda + \lambda(ps + 1 - p)\right)$$

$$= \exp\left(\lambda p(s - 1)\right)$$

and the result follows.   □

By its very construction, we see that

$$X(t) \stackrel{\text{dist.}}{=} \left(1 - e^{-t}\right) \circ \text{Poisson}(\rho)$$

and if $X(0)$ is not null, following the same reasoning, we have

$$X(t) \stackrel{\text{dist.}}{=} e^{-t} \circ X(0) + \left(1 - e^{-t}\right) \circ \text{Poisson}(\rho). \tag{5.10}$$

If $X(0)$ is distributed as a Poisson distribution of parameter ρ, then $X(t)$ is distributed as the sum of two independent Poisson random variables of the respective parameters ρe^{-t} and $\rho(1 - e^{-t})$; hence, $X(t)$ has the distribution of $X(0)$. We retrieve that the Poisson distribution of parameter ρ is the invariant and stationary measure of X.

5.4 Stochastic Analysis

5.4.1 Discrete Gradient and Divergence

Theorem 5.7 (Cameron–Martin Theorem) *Let N and N' be two Poisson point processes, with the respective intensities σ and σ'. Let us assume that $\sigma' \ll \sigma$, and let us denote $p = d\sigma'/d\sigma$. Moreover, if p belongs to $L^1(E \to \mathbf{R}; \sigma)$, then for every bounded function F, we have*

$$\mathbf{E}\left[F(N')\right] = \mathbf{E}\left[F(N) \exp\left(\int_E \ln p \, dN + \int_E (1 - p) d\sigma\right)\right].$$

Proof Step 1 We verify this identity for the exponential functions F of the form $\exp(-\int_E f dN)$. According to the definition [5.1],

$$\mathbf{E}\left[\exp\left(-\int_E f dN\right) \exp\left(\int_E \ln p \, dN + \int_E (1 - p) d\sigma\right)\right]$$

$$= \mathbf{E}\left[\exp\left(-\int_E (f - \ln p) dN\right)\right] \exp\left(\int_E (1 - p) d\sigma\right)$$

$$= \exp\left(-\int_E \left(1 - \exp\left(-f + \ln p\right)\right) d\sigma + \int_E (1 - p) d\sigma\right)$$

$$= \exp\left(-\int_E (1 - e^{-f}) p \, d\sigma\right)$$

$$= \mathbf{E}\left[F(N')\right].$$

Step 2 As a result, the measures on $\mathfrak{N}_E$, $\pi^\sigma_{N'}$, and $R d\pi^\sigma_N$ where

$$R = \exp\left(\int_E \ln p \, dN + \int_E (1 - p) d\sigma\right)$$

have the same Laplace transform. Therefore, in view of Theorem 5.1, they are equal, and the result follows for any bounded function F.

□

New Notations
In what follows, for a configuration ϕ

$$\phi \oplus x = \begin{cases} \phi, & \text{if } x \in \phi, \\ \phi \cup \{x\}, & \text{if } x \notin \phi. \end{cases}$$

(continued)

Similarly,

$$\phi \ominus x = \begin{cases} \phi \setminus \{x\}, & \text{if } x \in \phi, \\ \phi, & \text{if } x \notin \phi. \end{cases}$$

One of the essential formulas for the Poisson process is the following.

Theorem 5.8 (Campbell–Mecke Formula) *Let N be a Poisson process with intensity σ. For any random field $F : \mathfrak{N}_E \times E \to \mathbf{R}$ such that*

$$\mathbf{E}\left[\int_E |F(N, x)| d\sigma(x)\right] < \infty$$

then

$$\mathbf{E}\left[\int_E F(N \oplus x, x) \, d\sigma(x)\right] = \mathbf{E}\left[\int_E F(N, x) \, dN(x)\right]. \tag{5.11}$$

Proof **Step 1** According to the first definition of the Poisson process, for f with compact support and K a compact E, for any $t > 0$,

$$\mathbf{E}\left[\exp\left(-\int_E (f + \theta \mathbf{1}_K) dN\right)\right] = \exp\left(-\int_E 1 - e^{-f(x) - \theta \mathbf{1}_K(x)} d\sigma(x)\right).$$

According to the theorem of derivation under the summation sign, on one hand, we have

$$\frac{d}{d\theta} \mathbf{E}\left[\exp\left(-\int_E (f + \theta \mathbf{1}_K) \, dN\right)\right]\Big|_{\theta=0} = -\mathbf{E}\left[e^{-\int_E f dN} \int_E \mathbf{1}_K \, dN\right]$$

and on the other hand,

$$\frac{d}{d\theta} \exp\left(-\int_E 1 - e^{-f(x) - \theta \mathbf{1}_K(x)} d\sigma(x)\right)\Big|_{\theta=0}$$

$$= -\mathbf{E}\left[\int_E e^{-\int_E f dN + f(x)} \mathbf{1}_K(x) \, d\sigma(x)\right]. \tag{5.12}$$

As $\int_E f dN + f(x) = \int_E f d(N \oplus x)$, (5.11) is true for functions of the form $\mathbf{1}_K \otimes e^{-\int_E f dN}$.

Step 2 The measure

$$\mathfrak{C} : \mathcal{B}(\mathfrak{N}_E \times E) \longrightarrow \mathbf{R}^+$$

$$\Gamma \times K \longmapsto \mathbf{E}\left[\mathbf{1}_\Gamma(N)\int_E \mathbf{1}_K(x)\mathrm{d}N(x)\right]$$

is the so-called Campbell measure. If we consider the map

$$\mathfrak{T} : \mathfrak{N}_E \times E \longrightarrow \mathfrak{N}_E \times E$$

$$(\phi, x) \longmapsto (\phi \oplus x, x),$$

Equation (5.11) is equivalent to say that

$$\mathfrak{T}^*(\pi^\sigma \otimes \sigma) = \mathfrak{C}.$$

Moreover, (5.12) means that

$$\int_E e^{-\int_E f\mathrm{d}\phi}\mathbf{1}_K(x)\,\mathrm{d}\mathfrak{C}(\phi, x) = \int_E e^{-\int_E f\mathrm{d}\phi}\mathbf{1}_K(x)\,\mathrm{d}\mathfrak{T}^*(\pi^\sigma \otimes \sigma)(\phi, x).$$

Since a measure on $\mathfrak{N}_E$ is characterized by its Laplace transform, Eq. (5.12) is then sufficient to imply that (5.11) holds for any function F for which the two terms are meaningful.

$\square$

Definition 5.6 (Discrete Gradient) Let N be a Poisson process with intensity σ. Let $F : \mathfrak{N}_E \longrightarrow \mathbf{R}$ be a measurable function such that $\mathbf{E}\left[F(N)^2\right] < \infty$. We define Dom D as the set of square integrable random variables such that

$$\mathbf{E}\left[\int_E |F(N \oplus x) - F(N)|^2\mathrm{d}\sigma(x)\right] < \infty.$$

For $F \in$ Dom D, we set

$$D_x F(N) = F(N \oplus x) - F(N).$$

Example (Computation of $D_x F$) For example, for f deterministic belonging to $L^2(E \to \mathbf{R}; \sigma)$, $F = \int_E f\mathrm{d}N$ belongs to Dom D and $D_x F = f(x)$ because

$$F(N \oplus x) = \sum_{y\in N\cup\{x\}} f(y) = \sum_{y\in N} f(y) + f(x).$$

Similarly, if $F = \max_{y \in N} f(y)$, then

$$D_x F(N) = \begin{cases} 0 & \text{if } f(x) \leq F(N), \\ f(x) - F & \text{if } f(x) > F(N). \end{cases}$$

Definition 5.7 (Poisson Divergence) We denote by $\mathrm{Dom}_2\, \delta$, the set of vector fields such that

$$\mathbf{E}\left[\left(\int_E U(N \ominus x, x)\,(dN(x) - d\sigma(x))\right)^2\right] < \infty.$$

Then, for such vector fields U,

$$\delta U(N) = \int_E U(N \ominus x, x) dN(x) - \int_E U(N, x) d\sigma(x).$$

A consequence of Campbell–Mecke formula is the integration by parts formula.

Theorem 5.9 (Integration by Parts for Poisson Process) *For $F \in \mathrm{Dom}\, D$ and any $U \in \mathrm{Dom}_2\, \delta$,*

$$\mathbf{E}\left[\int_E D_x F(N)\, U(N, x) d\sigma(x)\right] = \mathbf{E}\left[F(N)\, \delta U(N)\right].$$

Proof By the very definition of D,

$$\mathbf{E}\left[\int_E D_x F(N)\, U(N, x)\, d\sigma(x)\right]$$

$$= \mathbf{E}\left[\int_E F(N \oplus x) U\big((N \ominus x) \oplus x,\, x\big) d\sigma(x)\right]$$

$$- \mathbf{E}\left[\int_E F(N) U(N, x)\, d\sigma(x)\right].$$

The Campbell–Mecke formula says that

$$\mathbf{E}\left[\int_E F(N \oplus x) U\big((N \ominus x) \oplus x,\, x\big) d\sigma(x)\right]$$

$$= \mathbf{E}\left[\int_E F(N) U(N \ominus x,\, x)\, dN(x)\right]. \tag{5.13}$$

Since we have assumed σ diffuse, for any $x \in E, \pi^{\sigma}\left(N(\{x\}) \geq 1\right) = 0$; hence,

$$U(N, x) = U(N \ominus x, x), \quad \pi^{\sigma} \otimes \sigma\text{-a.s.} \tag{5.14}$$

The result follows from the combination of (5.13) and (5.14).   □

Moreover, we have the analog to (2.34).

Corollary 5.1 *For any $U \in \mathrm{Dom}_2\, \delta$,*

$$\mathbf{E}\left[\delta U^2\right] = \mathbf{E}\left[\int_E U(N, x)^2 \, \mathrm{d}\sigma(x)\right]$$

$$+ \mathbf{E}\left[\int_E \int_E D_x U(N, y)\, D_y U(N, x)\, \mathrm{d}\sigma(x)\mathrm{d}\sigma(y)\right].$$

Proof We use the integration by parts formula to write

$$\mathbf{E}\left[\delta U^2\right] = \mathbf{E}\left[\int_E D_x \delta U \, U(N, x) \, \mathrm{d}\sigma(x)\right].$$

From the definition of D and δ,

$$D_x \delta U = \int_E U(N \ominus y \oplus x, y) \left(\mathrm{d}(N \oplus x)(y) - \mathrm{d}\sigma(y)\right)$$

$$- \int_E U(N \ominus y, y) \left(\mathrm{d}N(y) - \mathrm{d}\sigma(y)\right).$$

Recall the definition of the stochastic integral as a sum:

$$\int_E U\left(N \ominus y \oplus x, y\right) \mathrm{d}(N \oplus x)(y) = \sum_{y \in N \cup \{x\}} U\left(N \ominus y \oplus x, y\right)$$

$$= \sum_{y \in N} U\left(N \oplus x \ominus y, y\right) + U(N, x) = \int_E U\left(N \oplus x \ominus y, y\right) \mathrm{d}N(y) + U(N, x)$$

Hence, we get

$$D_x \delta U$$

$$= \int_E \left(U\left(N \oplus x \ominus y, y\right) - U(N \ominus y, y)\right) \left(\mathrm{d}N(y) - \mathrm{d}\sigma(y)\right) + U(N, x)$$

$$= \delta(D_x U) + U(N, x).$$

Thus,

$$\mathbf{E}\left[\delta U^2\right] = \mathbf{E}\left[\int_E \left(\delta(D_x U) + U(N, x)\right) U(N, x)\, d\sigma(x)\right]$$

$$= \mathbf{E}\left[\int_E U(N, x)^2 d\sigma(x)\right] + \int_E \mathbf{E}\left[\delta(D_x U)\, U(N, x)\right] d\sigma(x).$$

We may integrate by parts in the rightmost expectation, taking care to not mix the variables:

$$\mathbf{E}\left[\delta(D_x U)\, U(N, x)\right] = \mathbf{E}\left[\int_E D_x U(N, u)\, D_y U(N, x) d\sigma(y)\right].$$

This yields

$$\mathbf{E}\left[\delta U^2\right] = \mathbf{E}\left[\int_E U(N, x)^2 d\sigma(x)\right] + \mathbf{E}\left[\int_E D_x U(N, u)\, D_y U(N, x) d\sigma(y)\right],$$

hence the result.   □

5.4.2 Functional Calculus

The Glauber point process, denoted by $\mathcal{G}$, is a Markov process with values in $\mathfrak{N}_E$ whose stationary and invariance measure is π^σ. Its generator is $\mathcal{L} = -\delta D$. Its semigroup satisfies a Mehler-like description. It is the key stone of the Dirichlet structure associated to π^σ.

Definition 5.8 The Markov process $\mathcal{G}$ is constructed as follows:

- $\mathcal{G}(0) = \phi \in \mathfrak{N}_E$.
- Each atom of ϕ has a life duration, independent of that of the other atoms, exponentially distributed with parameter 1.
- Atoms are born at moments following a Poisson process on the half-line, with intensity $\sigma(E)$. On its appearance, each atom is localized independently from all the others according to $\sigma/\sigma(E)$. It is also assigned in an independent manner, a life duration exponentially distributed with parameter 1.

At every instant, $\mathcal{G}(t)$ is a configuration of E, see Fig. 5.2. We first observe that the total number of atoms of $\mathcal{G}(t)$ follows exactly the same dynamics as the number of busy servers in a M/M/∞ queue with parameters $\sigma(E)$ and 1.

Theorem 5.10 (Glauber Process) *For any $t > 0$, the process $\mathcal{G}(t)$ has the distribution of*

$$e^{-t} \circ \mathcal{G}(0) \oplus (1 - e^{-t}) \circ N' \tag{5.15}$$

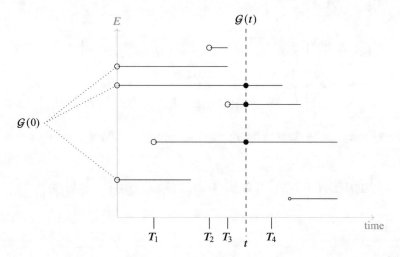

Fig. 5.2 Realization of a trajectory of $\mathcal{G}$.

where N' is an independent copy of N.

Assume that $\mathcal{G}(0)$ is a point Poisson process with intensity σ. Then, $\mathcal{G}(t)$ has the distribution of N for any t.

Proof We can separate the atoms of $\mathcal{G}$ in two sets: $\mathcal{G}^o$ is the set of particles that were present at the origin and are still alive and $\mathcal{G}^\dagger$ is the set of fresh particles that were born after time 0 and are still alive. By construction, these two sets are independent.

Moreover, the particles of $\mathcal{G}^o$ alive at t correspond to an e^{-t}-thinning of the original configuration; thus

$$\mathcal{G}^o(t) \overset{\text{dist.}}{=} e^{-t} \circ \mathcal{G}(0). \tag{5.16}$$

For two disjoint parts A and B of E, by construction, the atoms of $\mathcal{G}^\dagger$ that belong to A (respectively, B) appear as a $\mathbf{1}_A$-thinning (respectively, $\mathbf{1}_B$-thinning) of the Poisson process that represents the birth dates. Then, Theorem 5.6 says that the date of birth $\mathcal{G}^\dagger \cap A$ is a Poisson point process of intensity $\sigma(A)$, independent of $\mathcal{G}^\dagger \cap B$.

Following the computations made for the M/M/∞ queue, we see that

$$(\mathcal{G}^\dagger \cap A)(t) \overset{\text{dist}}{=} \text{Poisson}\big((1 - e^{-t})\sigma(A)\big).$$

Hence, according to Theorem 5.4,

$$\mathcal{G}^\dagger \cap A \overset{\text{dist}}{=} (1 - e^{-t}) \circ (N' \cap A). \tag{5.17}$$

Then, (5.15) follows from (5.16)and (5.17).

If $G(0)$ is distributed as N, then Theorem 5.6 entails that $e^{-t} \circ G(0)$ is a Poisson process of intensity $e^{-t}\sigma$. Thus, the superposition theorem 5.5 implies that G has the distribution of N. □

As all the sojourn times are exponentially distributed, G is a Markov process with values in $\mathfrak{N}_E$. Far from the idea of developing the general theory of Markov processes in the space of measures, we can study its infinitesimal generator and its semi-group. Equation (5.17) means that we have the Poisson–Mehler formula.

Theorem 5.11 *For any $t \geq 0$, for $F \in L^1(\mathfrak{N}_E \to \mathbf{R}; \pi^\sigma)$:*

$$\mathcal{P}_t(\phi) := \mathbf{E}\left[F\big(G(t)\big) \mid G(0) = \phi\right] = \mathbf{E}\left[F\big(e^{-t} \circ \phi \oplus (1 - e^{-t}) \circ N'\big)\right] \quad (5.18)$$

where the expectation is taken with respect to the law of N'.

Theorem 5.12 *The infinitesimal generator of G, denoted by $\mathcal{L}$, is given by*

$$-\mathcal{L}F(\phi) = \int_E \Big(F(\phi \oplus x) - F(\phi)\Big) d\sigma(x)$$

$$+ \int \Big(F(\phi \ominus x) - F(\phi)\Big) d\phi(x) \quad (5.19)$$

for F bounded from $\mathfrak{N}_E$ into $\mathbf{R}$.

Proof At time t, there may be either a death or a birth. At a death time, we choose the atom to kill uniformly among the existing ones, so that each atom has a probability $\phi(E)^{-1}$ of being killed. Since all atoms have a lifetime that follows a unit exponential distribution, the death rate is $\phi(E)$. Therefore, the transition from ϕ to $\phi \ominus x$ takes place at rates of 1 for any $x \in \phi$.

The birth rate is $\sigma(E)$, and the position of the new atom is distributed according to the measure $\sigma/\sigma(E)$, so the transition from ϕ to $\phi \oplus x$ occurs at a rate $d\sigma(x)$ for each $x \in E$. From this reasoning, we deduce (5.19). □

Theorem 5.13 (Ergodicity) *The semi-group $\mathcal{P}$ is ergodic. Moreover, $\mathcal{L}$ is invertible from L_0^2 in $L_0^2 = L^2(\mathfrak{N}_E \to \mathbf{R}; \pi^\sigma) \cap \{F, \mathbf{E}[F] = 0\}$, and we have*

$$\mathcal{L}^{-1}F = \int_0^\infty \mathcal{P}_t F \, dt. \quad (5.20)$$

For any $x \in E$ and any $t > 0$,

$$D_x \mathcal{P}_t F = e^{-t} \mathcal{P}_t D_x F. \quad (5.21)$$

If, in addition, F is such that

$$\sup_{\phi \in \mathfrak{N}_E} \int_E |D_x F(\phi)|^2 d\sigma(x) < \infty,$$

then, with probability 1, we have

$$\int_E \left| D_x \left(\mathcal{L}^{-1} F(\phi) \right) \right|^2 d\sigma(x) \leq \sup_{\phi \in \mathfrak{N}_E} \int_E |D_x F(\phi)|^2 d\sigma(x). \tag{5.22}$$

***Proof* Step 1** By dominated convergence, we deduce from (5.18) that

$$\mathcal{P}_t F(\phi) \xrightarrow{t \to \infty} \mathbf{E}[F(N)],$$

that is to say, $\mathcal{P}$ is ergodic.

Step 2 The property (5.20) is a well known relation between the semi-group and infinitesimal generator. Formally, without worrying about the convergence of the integrals, we have

$$\mathcal{L} \left(\int_0^\infty \mathcal{P}_t F \, dt \right) = \int_0^\infty \mathcal{L} \mathcal{P}_t F \, dt$$

$$= -\int_0^\infty \frac{d}{dt} \mathcal{P}_t F \, dt$$

$$= F - \mathbf{E}[F] = F$$

according to ergodicity of $\mathcal{P}$ and as F is centered.

Step 3 Starting from the formula (5.18),

$$D_x \mathcal{P} F(t) = \mathbf{E} \left[F \left(e^{-t} \circ (\phi \oplus x) \oplus (1 - e^{-t}) \circ N' \right) \right]$$

$$- \mathbf{E} \left[F \left(e^{-t} \circ \phi \oplus (1 - e^{-t}) \circ N' \right) \right].$$

Since the thinning operation is distributive on the superposition of point processes, we get

$$D_x \mathcal{P} F(t) = \mathbf{E} \left[F \left(e^{-t} \circ \phi \oplus e^{-t} \circ x \oplus (1 - e^{-t}) \circ N' \right) \right]$$

$$- \mathbf{E} \left[F \left(e^{-t} \circ \phi \oplus (1 - e^{-t}) \circ N' \right) \right].$$

At time t, either $e^{-t} \circ x = x$ or $e^{-t} \circ x = \emptyset$; the former event appears with probability e^{-t} and the latter with the complementary probability; hence,

$$
\mathbf{E}\left[F\left(e^{-t} \circ \phi \oplus e^{-t} \circ x \oplus (1 - e^{-t}) \circ N'\right)\right]
$$
$$
= e^{-t}\,\mathbf{E}\left[F\left(e^{-t} \circ \phi \oplus x \oplus (1 - e^{-t}) \circ N'\right)\right]
$$
$$
+ (1 - e^{-t})\,\mathbf{E}\left[F\left(e^{-t} \circ \phi \oplus (1 - e^{-t}) \circ N'\right)\right].
$$

It follows that

$$
D_x \mathcal{P} F(t) = e^{-t}\,\mathbf{E}\left[F\left(e^{-t} \circ \phi \oplus x \oplus (1 - e^{-t}) \circ N'\right)\right]
$$
$$
- e^{-t}\,\mathbf{E}\left[F\left(e^{-t} \circ \phi \oplus (1 - e^{-t}) \circ N'\right)\right] = e^{-t}\mathcal{P} D_x F(t).
$$

Step 4   As a consequence of the previous part of this proof,

$$
\int_E \left|D_x\big(\mathcal{L}^{-1} F(N)\big)\right|^2 d\sigma(x) = \int_E \left(\int_0^\infty e^{-t} \mathcal{P}_t D_x F(N) dt\right)^2 d\sigma(x).
$$

According to the Jensen formula, we get

$$
\int_E \left|D_x\big(\mathcal{L}^{-1} F(N)\big)\right|^2 d\sigma(x) \le \int_E \int_0^\infty e^{-t}\left|\mathcal{P}_t D_x F(N)\right|^2 dt\, d\sigma(x).
$$

The representation (5.18) and Jensen inequality imply that $|\mathcal{P}_t G|^2 \le \mathcal{P}_t G^2$; thus,

$$
\int_E \left|D_x\big(\mathcal{L}^{-1} F(N)\big)\right|^2 d\sigma(x)
$$
$$
\le \int_E \int_0^\infty e^{-t}\,\mathbf{E}\left[(D_x F)^2\big(e^{-t} \circ N \oplus (1 - e^{-t}) \circ N'\big) \mid N\right] dt\, d\sigma(x)
$$
$$
= \int_0^\infty e^{-t} \int_E \mathbf{E}\left[(D_x F)^2\big(e^{-t} \circ N \oplus (1 - e^{-t}) \circ N'\big) \mid N\right] d\sigma(x) dt
$$
$$
= \int_0^\infty e^{-t} \sup_{\phi \in \mathfrak{N}_E} \int_E (D_x F)^2(\phi) d\sigma(x)\, dt
$$
$$
= \sup_{\phi \in \mathfrak{N}_E} \int_E (D_x F)^2(\phi)\, d\sigma(x).
$$

The proof is thus complete.

$\square$

Theorem 5.14 (Covariance Identity)   *Let F and G be two functions belonging to* Dom *D. The following identity is satisfied:*

$$\mathbf{E}\left[\int_E D_x F(N)\, D_x G(N) \mathrm{d}\sigma(x)\right] = \mathbf{E}\left[F(N)\,\mathcal{L}G(N)\right].$$

In particular, if G is centered

$$\mathbf{E}\left[F(N)G(N)\right] = \mathbf{E}\left[\int_E D_x F(N)\, D_x\left(\mathcal{L}^{-1}G\right)(N)\, \mathrm{d}\sigma(x)\right]. \tag{5.23}$$

Proof Let *F* and *G* belong to Dom *D*. We are going to show the most important formula

$$\mathcal{L} = \delta D. \tag{5.24}$$

By definition,

$$\delta D F(N) = \int_E D_x F(N \ominus x)\, \mathrm{d}N(x) - \int_E D_x F(N)\, \mathrm{d}\sigma(x)$$

$$= \int_E \left(F(N) - F(N \ominus x)\right) \mathrm{d}N(x) - \int_E \left(F(N \oplus x) - F(N)\right) \mathrm{d}\sigma(x). \tag{5.25}$$

It remains to compare (5.25) and (5.19).   □

Concentration Inequality: Scheme of Proof

The proof of the concentration inequality follows a classical scheme that can be applied to Wiener functionals as well. Considering that X is a centered random variable and $r > 0$, the well known trick is to use the Markov inequality in a subtle manner:

$$\mathbf{P}(X > r) = \mathbf{P}(e^{\theta X} > e^{\theta r}) \le e^{-\theta r}\mathbf{E}\left[e^{\theta X}\right]. \tag{5.26}$$

The goal is then to find a somehow explicit bound of $\mathbf{E}\left[e^{\theta X}\right]$ and optimize the right-hand side of (5.26) with respect to θ.

The computation of the upper-bound of $\mathbf{E}\left[e^{\theta X}\right]$ relies on the Herbst principle. Compute

$$\frac{\mathrm{d}}{\mathrm{d}\theta}\mathbf{E}\left[e^{\theta X}\right] = \mathbf{E}\left[X\, e^{\theta X}\right]$$

(continued)

and do whatever it costs to bound it by something of the form

$$\mathbf{E}\left[X\,e^{\theta X}\right] \leq (\text{function of } \theta) \times \mathbf{E}\left[e^{\theta X}\right].$$

This amounts to bound the logarithmic derivative of $\mathbf{E}\left[e^{\theta X}\right]$. It remains to integrate this last inequality to obtain the desired bound. The difficulty here is that $e^{\theta X}$ appears on both sides of the inequality. This means that we can only use $L^1 - L^\infty$ inequalities, hence, the stringent conditions on the sup norms that will appear.

Theorem 5.15 (Concentration Inequality) *Let N be a Poisson process with intensity σ on E. Let $F : \mathfrak{N}_E \to \mathbf{R}$ such that*

$$D_x F(N) \leq \beta, \ (\sigma \otimes \pi^\sigma) - a.e. \text{ and } \sup_{\phi \in \mathfrak{N}_E} \int_E |D_x F(\phi)|^2 \, d\sigma(x) \leq \alpha^2, \ \pi^\sigma - a.e.$$

For any $r > 0$, we have the following inequality:

$$\pi^\sigma\left(F(N) - \mathbf{E}[F(N)] > r\right) \leq \exp\left(-\frac{r}{2\beta} \ln(1 + \frac{r\beta}{\alpha^2})\right).$$

Proof **Step 1** As a preliminary computation, remark that

$$D_x e^{\theta F(N)} = e^{\theta F(N \oplus x)} - e^{\theta F(N)}$$

$$= \left(e^{\theta D_x F(N)} - 1\right) e^{\theta F(N)}. \tag{5.27}$$

Step 2 Let F be a bounded function of null expectation. According to Theorem 5.14 and (5.27), we can write the following identities:

$$\mathbf{E}\left[F(N)\,e^{\theta F(N)}\right] = \mathbf{E}\left[\int D_x\left(\mathcal{L}^{-1}F(N)\right) D_x\left(e^{\theta F(N)}\right) d\sigma(x)\right]$$

$$= \mathbf{E}\left[\int_E D_x\left(\mathcal{L}^{-1}F(N)\right) \left(e^{\theta D_x F(N)} - 1\right) e^{\theta F(N)} d\sigma(x)\right].$$

Step 3 We want to benefit from the fact that the function $\Psi : (x \mapsto (e^x - 1)/x)$ is continuously increasing on $\mathbf{R}$; therefore, we impose its presence

$$\mathbf{E}\left[F(N)e^{\theta F(N)}\right]$$

$$= \theta\,\mathbf{E}\left[\int_E D_x\left(\mathcal{L}^{-1}F(N)\right)D_x F(N)\,\Psi\left(\theta D_x F(N)\right)e^{\theta F(N)}\,d\sigma(x)\right].$$

Since $D_x F \leq \beta$, we obtain

$$\left|\mathbf{E}\left[F(N)e^{\theta F(N)}\right]\right|$$

$$\leq \theta\,\Psi(\theta\beta)\,\mathbf{E}_\sigma\left[e^{\theta F(N)}\int_E D_x\left(\mathcal{L}^{-1}F(N)\right)D_x F(N)\,d\sigma(x)\right].$$

Use the Cauchy–Schwarz inequality to get

$$\left|\int_E D_x\left(\mathcal{L}^{-1}F(N)\right)D_x F(N)\,d\sigma(x)\right|$$

$$\leq \left|\int_E D_x\left(\mathcal{L}^{-1}F(N)\right)^2 d\sigma(x)\right|^{1/2} \times \left|\int_E D_x F(N)^2\,d\sigma(x)\right|^{1/2} = A_1 \times A_2.$$

According to the hypothesis, $A_2 \leq \alpha$ and Eq. (5.22) tells that so does A_1. This implies that

$$\frac{d}{d\theta}\log\mathbf{E}\left[e^{\theta F(N)}\right] \leq \alpha^2 \frac{e^{\theta\beta} - 1}{\beta}.$$

Therefore,

$$\mathbf{E}\left[e^{\theta F(N)}\right] \leq \exp\left(\frac{\alpha^2}{\beta}\int_0^\theta (e^{\beta u} - 1)du\right).$$

For $x > 0$, for any $\theta > 0$,

$$\pi^\sigma\left(F(N) > x\right) = \pi^\sigma\left(e^{\theta F(N)} > e^{\theta x}\right)$$

$$\leq e^{-\theta x}\mathbf{E}\left[e^{\theta F(N)}\right]$$

$$\leq e^{-\theta x}\exp\left(\frac{\alpha^2}{\beta}\int_0^\theta (e^{\beta u} - 1)\,du\right). \tag{5.28}$$

This result is true for any θ, so we can optimize with respect to θ. At fixed x, we search the value of θ that cancels the derivative of the right-hand side with respect to θ. Plugging this value into (5.28), we can obtain the result.

$\square$

5.5 A Quick Refresher About the Poisson Process on the Line

The Poisson process on the real line is a particular case of the Poisson point process defined above. It admits a more convenient definition based on a sequence of independent exponential random variables.

Definition 5.9 Consider $(\xi_n, \ n \geq 1)$ a sequence of independent random variables sharing the same exponential distribution of parameter λ. Consider

$$T_n = \sum_{i=1}^{n} \xi_i.$$

A point process N on $E = \mathbf{R}^+$ of intensity λ is the point process whose atoms are $(T_n, \ n \geq 1)$.

To be compatible with the usual notations of time indexed processes, we set

$$N(t) = N([0, t]) = \sum_{i=1}^{\infty} \mathbf{1}_{\{T_i \leq t\}}.$$

With the vocabulary of this chapter, N is a Poisson point process of intensity measure $\sigma = \lambda \ell$. Following Theorem 5.4, N is a time indexed process with independent and stationary increments. The properties of superposition and thinning are definitely valid for N. Since we have a notion of time, we can define the filtration $\mathcal{F}_t^N = \sigma(N(s), \ u \leq t)$. The additional property is that, on any time interval $[0, T]$, the process

$$\tilde{N} : t \longmapsto N(t) - \lambda t$$

is a martingale of square bracket

$$\left\langle \tilde{N} \right\rangle_t = \lambda t.$$

For any $\mathcal{F}^N$-adapted, left-continuous process $u \in L^2(\mathfrak{N}_{[0,T]} \times [0, T] \to \mathbf{R}; \pi^\sigma \otimes \sigma)$, the *compensated integral* with respect to N is the martingale

$$\int_0^t u(N, s) \, d\tilde{N}(s) := \int_0^t u(N, s) \, dN(s) - \int_0^t u(N, s) \, \lambda \, ds,$$

of square bracket

$$t \longmapsto \int_0^t u(s)^2 \lambda \, ds.$$

This means that we have the Itô isometry formula for Poisson integrals:

$$\mathbf{E}\left[\left(\int_0^t u(N, s) \, d\tilde{N}(s)\right)^2\right] = \mathbf{E}\left[\int_0^t u(s)^2 \, \lambda \, ds\right]. \tag{5.29}$$

When u is adapted and left-continuous, $u(N, s)$ depends on the trajectory of N until time s^-; hence, if a sample-path of N is modified after time s, this does not change the value of $u(N, s)$. More precisely, we have

$$u(N \ominus t, \, s) = u(N, s) \text{ for any } t \geq s.$$

It follows that the Poisson divergence coincides with the compensated integral, and Corollary 5.1 is an extension of the Itô isometry (5.29).

5.6 Problems

5.1 (Chaos Decomposition for Poisson Functionals) For $f \in L^2(E \to \mathbf{R}; \sigma)$, let

$$\Lambda_f(N) = \exp\left(-\int_E f \, dN + \int_E (1 - e^{-f}) \, d\sigma\right).$$

We already know from Theorem 5.7 that $\mathbf{E}[\Lambda_f] = 1$. For a configuration ϕ, we introduce its factorial moment measure of order $k \geq 1$:

$$\phi^{(k)}(A) = \int \mathbf{1}_A(x_1, \cdots, x_k) \, d(\mu \ominus \oplus_{j=1}^{k-1} x_j)(x_k) \, d(\mu \ominus \oplus_{j=1}^{k-2} x_j)(x_{k-1})$$

$$\cdots d(\mu \ominus x_1)(x_2) \, d\mu(x_1).$$

We set $N^{(0)}(f) = 1$.

1. Show that

$$D_x \Lambda_f(N) = \Lambda_f(N)(e^{-f(x)} - 1)$$

and that

$$\mathbf{E}\left[D^{(n)}_{x_1 \ldots x_n} \Lambda_f(N)\right] = \prod_{j=1}^{n} \left(e^{-f(x_j)} - 1\right).$$

2. Show that

$$\delta^{(2)}(f^{\otimes(2)}) = N^{(2)}(f \otimes f) - 2N(f)\sigma(f) + \sigma(f)^2$$

and more generally that

$$\delta^{(n)}(f^{\otimes(n)}) = \sum_{j=0}^{n} \binom{n}{j}(-1)^{n-j} N^{(j)}(f^{\otimes(j)}) \left(\int_E f \, d\sigma\right)^{n-j}.$$

3. Set $\delta^{(0)}(f^{\otimes(0)}) = 1$. Show that

$$\sum_{n=0}^{N(E)} \frac{1}{n!} \delta^{(n)}\left((e^{-f} - 1)^{\otimes(n)}\right)$$

$$= \exp\left(-\int_E (e^{-f} - 1) \, d\sigma\right) \sum_{j=0}^{\infty} \frac{1}{j!} N^{(j)}((e^{-f} - 1)^{\otimes(j)}).$$

4. If $X_1, \cdots, X_{N(E)}$ are the atoms of N, show that

$$\sum_{j=0}^{\infty} \frac{1}{j!} N^{(j)}\left((e^{-f} - 1)^{\otimes(j)}\right) = \sum_{J \subset \{1,2,\cdots,N(E)\}} \prod_{i \in J} (e^{-f(X_i)} - 1)$$

$$= \prod_{j=1}^{N(E)} e^{-f(X_j)} = e^{-\int_E f \, dN}.$$

Hence, provided that we show the convergence of the sums in $L^2(\mathfrak{N}_2 \to \mathbf{R}; \pi^\sigma)$, we have proved that

$$\Lambda_f(N) = 1 + \sum_{n=1}^{\infty} \delta^{(n)}\left(\mathbf{E}\left[D^{(n)}\Lambda_f(N)\right]\right).$$

Taking for granted that the vector space spanned by the Λ_f's when f goes through $L^2(E \to \mathbf{R}; \sigma)$ is dense in $L^2(\mathfrak{N}_2 \to \mathbf{R}; \pi^\sigma)$, we obtain the chaos decomposition for functionals of Poisson process:

$$F = \mathbf{E}[F] + \sum_{n=1}^{\infty} \frac{1}{n!} \delta^{(n)} \left(\mathbf{E}\left[D^{(n)} F(N) \right] \right). \qquad (5.30)$$

5.7   Notes and Comments

The interested reader could find more details about the topology of configuration spaces in [3]. The construction of the Poisson point process in more general spaces than Polish spaces can be found in [4]. For more information about the Malliavin calculus for Poisson process, see [4, 5]. The construction of the Glauber process follows [2], but the presentation given here emphasizes the invariance property of the Poisson process: N is a Poisson point process if and only if

$$N \stackrel{\text{dist}}{=} p \circ N' \oplus (1 - p) \circ N''$$

where N' and N'' are independent copies of the point process N. The concentration inequality has already been published in [1]. For an alternative proof, see [6]. As for the Brownian motion, (5.30) can be the starting point of the definition of the operators D and δ, see [5].

Reference

1. L. Decreusefond, P. Moyal, *Stochastic Modeling and Analysis of Telecom Networks* (ISTE Ltd/Wiley, London/Hoboken, 2012)
2. L. Decreusefond, M. Schulte, C. Thäle, Functional Poisson approximation in Kantorovich–Rubinstein distance with applications to U-statistics and stochastic geometry. Ann. Probab. **44**(3), 2147–2197 (2016)
3. O. Kallenberg, *Random Measures, Theory and Applications*. Probability Theory and Stochastic Modelling, vol. 77 (Springer, Cham, 2017)
4. G. Last, M. Penrose, *Lectures on the Poisson Process*. Institute of Mathematical Statistics Textbooks, vol. 7 (Cambridge University Press, Cambridge, 2018)
5. N. Privault, *Stochastic Analysis in Discrete and Continuous Settings with Normal Martingales*. Lecture Notes in Mathematics, vol. 1982 (Springer, Berlin, 2009)
6. L. Wu, A new modified logarithmic Sobolev inequality for Poisson point processes and several applications. Probab. Theory Related Fields **118**(3), 427–438 (2000)

Chapter 6
The Malliavin–Stein Method

Abstract The Stein's method, initiated in the 1970s by Charles Stein, is a procedure to estimate the rate of convergence in CLT-like theorems. It gained a new momentum in the beginning of the millennium thanks to the insights given by the Malliavin calculus.

6.1 Principle

Among the more than seventy known distances between probability measures, the most classical ones are the Prokhorov–Lévy and Fortet–Mourier (or Bounded Lipschitz) distances.

Definition 6.1 For μ and ν two probability measures on a metric space (E, d) with Borelean σ-field $\mathcal{A}$, the Prokhorov–Lévy is defined as

$$\text{dist}_{\text{PL}}(\mu, \nu) = \max \left(\inf \{\epsilon; \mu(A) \le \nu(A^\epsilon) + \epsilon, \text{ for all closed } A \subset E\}, \right.$$
$$\left. \inf \{\epsilon; \nu(A) \le \mu(A^\epsilon) + \epsilon, \text{ for all closed } A \subset E\} \right)$$

where $A^\epsilon = \{x, d(x, A) \le \epsilon\}$. The Fortet–Mourier distance is defined as

$$\text{dist}_{\text{FM}}(\mu, \nu) = \sup_{f \in \text{BL}} \left(\int_E f \, d\mu - \int_E f \, d\nu \right)$$

where BL is the set of bounded Lipschitz functions:

$$\text{BL} = \left\{ f : E \to \mathbf{R}, f \text{ is bounded by } 1 \text{ and } |f(x) - f(y)| \le d(x, y), \forall x, y \in E \right\}.$$

Theorem 6.1 *If* $\text{dist}_{\text{PL}}(\mu, \nu) \le 1$ *or* $\text{dist}_{\text{FM}}(\mu, \nu) \le 2/3$,

$$\frac{2}{3} \text{dist}_{\text{PL}}(\mu, \nu)^2 \le \text{dist}_{\text{FM}}(\mu, \nu).$$

L. Decreusefond, *Selected Topics in Malliavin Calculus*,
Bocconi & Springer Series 10, https://doi.org/10.1007/978-3-031-01311-9_6

If (E, d) is separable, then

$$\text{dist}_{FM}(\mu, \nu) \leq 2 \, \text{dist}_{PL}(\mu, \nu).$$

Thus, if (E, d) is separable, the two metrics define the same topology on the set of probability measures on E. Furthermore, the three following properties are equivalent:

1. *$\int_E f d\mu_n \to \int_E f d\mu$ for all bounded and continuous functions from E to $\mathbf{R}$.*
2. *$\text{dist}_{FM}(\mu_n, \mu) \to 0$.*
3. *$\text{dist}_{PL}(\mu_n, \mu) \to 0$.*

Another class of distances between probability measures is given by the optimal transportation problem:

$$\mathfrak{D}_c(\mu, \nu) = \inf_{\gamma \in \Sigma_{\mu,\nu}} \int_{E \times E} c(x, y) d\gamma(x, y)$$

where $c : E \times E \to \mathbf{R}^+ \cup \{+\infty\}$ is a lower semi-continuous *cost* function and $\Sigma_{\mu,\nu}$ is the set of probability measures whose first marginal is μ and second marginal is ν. When c is a distance on E, $\mathfrak{D}_c$ defines the so-called Kantorovich–Rubinstein or Wasserstein-1 distance between probability measures on E. It admits an alternative characterization very similar to the Fortet–Mourier distance.

Theorem 6.2 *For c a distance on the metric space (E, d),*

$$\mathfrak{D}_c(\mu, \nu) = \sup_{f \in \text{Lip}_1(E,c)} \left(\int_E f d\mu - \int_E f d\nu \right) \tag{6.1}$$

where $\text{Lip}_1(E, c)$ is the set of Lipschitz functions

$$\text{Lip}_1(E, c) = \left\{ f : E \to \mathbf{R}, |f(x) - f(y)| \leq c(x, y), \forall x, y \in E \right\}.$$

We denote $\mathfrak{D}_c$ by dist_{KR}.

Remark 6.1 Remark that we do not need to put an absolute value in the right-hand side of (6.1) since $(-f) \in \text{Lip}_1(E, c)$ as soon as f is Lipschitz.

We also have:

Theorem 6.3 *The following two properties are equivalent:*

1. *$\text{dist}_{KR}(\mu_n, \mu) \to 0$.*
2. *$\int_E f d\mu_n \to \int_E f d\mu$ for all bounded and continuous functions from E to $\mathbf{R}$ and for some (and then for all) $x_0 \in E$*

$$\int_E c(x_0, x) d\mu_n(x) \xrightarrow{n \to \infty} \int_E c(x_0, x) d\mu(x).$$

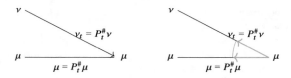

Fig. 6.1 Construct a transformation of the measures that leaves μ invariant and ultimately transforms ν into μ (left). Then, reverse time and control the gradient of the transformation (right)

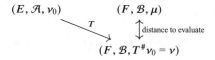

Fig. 6.2 The global scheme of the Stein method

These two examples give raise to several distances of the same form

$$\operatorname{dist}_{\mathfrak{F}}(\mu, \nu) = \sup_{f \in \mathfrak{F}} \left(\int_E f \, d\mu - \int_E f \, d\nu \right)$$

where $\mathfrak{F}$ is a space of test functions. For instance, if $\mathfrak{F}$ is the set of indicator functions of intervals, we retrieve the Kolmogorov distance. The Stein method is particularly well suited to estimate such quantities. For technical reasons, it is often necessary to consider sets of test functions smaller than $\operatorname{Lip}_1(E, c)$ even if we lose the nice equivalence with convergence in distribution.

The abstract description of the Stein method is to construct an homotopy between the two measures μ and ν and then control the distance between μ and ν by controlling the gradient of the homotopy (Fig. 6.1).

More precisely, the basic setting consists of a *target distribution* μ to which we will compare a distribution ν. The probability measure μ lives on a metric space $(E, \mathcal{A})$, and ν is often defined as the image measure of a measure ν_0 on $(F, \mathcal{B})$ by a map $T : F \to E$ (Fig. 6.2).

For instance, if we want to evaluate the rate of convergence in the law of rare events, we take $E = \mathbf{N}$, $F = \{0, \cdots, n\}$, ν_0 is the binomial distribution of parameters $(n, p/n)$ and T is the embedding from F into E. For the usual Central Limit Theorem, $E = \mathbf{R}$ and μ is the standard Gaussian distribution, $F = \mathbf{R}^n$ and $\nu_0 = \rho^{\otimes n}$ where ρ is the common distribution of the X_i's assumed to be centered and of unit variance. We take $T(x_1, \cdots, x_n) = n^{-1/2} \sum_{i=1}^{n} x_i$.

By comparison, the Skorokhod embedding method consists in finding S_1 and S_2 such that S_1 maps F into E and $S_2 : F \to E$ such that the image measure of μ by S_2 is ν_0. We then compare the distance between the realizations $S_1(\omega)$ and $S_2(\omega)$ in E. All the difficulty of this method is to devise the coupling between μ and ν_0, i.e., to find the convenient map S_2, see Fig. 6.3.

The Malliavin–Stein method assumes that there exists a Dirichlet structure on $(E, \mathcal{A}, \nu_0)$ and on $(F, \mathcal{B}, \mu)$.

Fig. 6.3 The global scheme
of the Skorokhod embedding
method

$$(E, \mathcal{A} \xleftarrow[S_1]{} (F, \mathcal{B}, \mu)$$

$$\uparrow_{\text{dist.}}$$

$$(E, \mathcal{A}, \nu_0) \xleftarrow[S_2]{} (F, \mathcal{B}, \mu)$$

Definition 6.2 A Dirichlet structure on $(E, \mathcal{A}, \nu_0)$ is a set of four elements X^0, L^0, $(P_t^0, t \geq 0)$, $\mathcal{E}^0$, where X^0 is a strong Feller process with values in E whose generator is L^0 and its semi-group is P^0: for $f : E \to \mathbf{R}$ sufficiently regular

$$P_t^0 f(x) = \mathbf{E}[f(X(t)) \mid X(0) = x]$$

$$\frac{\mathrm{d}}{\mathrm{d}t} P_t^0 f(x) = L^0 P_t^0 f(x)$$

$$= P_t^0 L^0 f(x).$$

Furthermore, ν_0 is the stationary and invariant distribution of X^0, and the Dirichlet form is defined by

$$\mathcal{E}^0(f, g) = \frac{\mathrm{d}}{\mathrm{d}t} \int_E P_t f(x) g(x) \mathrm{d}\nu_0(x) \bigg|_{t=0}.$$

Remark 6.2 We will not dwell into the theory of Dirichlet forms, but it must be noted that given one of the elements of the quadruple, one can construct, at least in an abstract way, the three other elements.

Remark 6.3 Actually, we do not really need $\mathcal{E}^0$ but rather the *carré du champ* operator defined by

$$\Gamma^0(f, g) = \frac{1}{2} \left(L^0(fg) - f L^0 g - g L^0 f \right), \qquad (6.2)$$

which is such that

$$\mathcal{E}^0(f, g) = \int_E \Gamma^0(f(x), g(x)) \, \mathrm{d}\nu_0(x).$$

In this setting, the most important formula is again an avatar of the integration by parts formula:

Theorem 6.4 *For f and g in $\mathrm{Dom}_2 L^0$ (i.e., such that $f \in L^2(E \to \mathbf{R}; \nu_0)$, $L^0 f$ is well defined and belongs to $L^2(E \to \mathbf{R}; \nu_0)$),*

$$\mathbf{E}\left[\Gamma^0(f, g)\right] = -\mathbf{E}\left[f L^0 g\right]. \qquad (6.3)$$

Proof We note that $P_t^0 1 = 1$; hence, we have $L^0 1 = 0$. Furthermore, since L^0 is self-adjoint

$$\mathbf{E}\left[L^0 f\right] = 0$$

for any $f \in \mathrm{Dom}\, L^0$. Thus, (6.2) yields

$$\mathbf{E}\left[\Gamma^0(f, g)\right] = -\frac{1}{2}\left(\mathbf{E}\left[f\, L^0 g\right] + \mathbf{E}\left[g\, L^0 f\right]\right)$$
$$= -\frac{1}{2}\left(\mathbf{E}\left[f\, L^0 g\right] + \mathbf{E}\left[f\, L^0 g\right]\right)$$

by the self-adjointness of L^0. $\qquad\square$

*Example (Gaussian Measure on **R**)* If we look for a Markov process with values in $E = \mathbf{R}$ whose stationary distribution is the standard Gaussian measure on $\mathbf{R}$ denoted by v, we may think of the Ornstein–Uhlenbeck process: It can be defined as the solution of the stochastic differential equation

$$X(t, x) = x - \int_0^t X(s, x)\mathrm{d}s + \sqrt{2}\, B(t) \tag{6.4}$$

where B is a standard Brownian motion. We can also write

$$X(t, x) = e^{-t}x + \sqrt{2} \int_0^t e^{-(t-s)}\mathrm{d}B(s)$$

so that

$$X(t, x) \sim \mathcal{N}(e^{-t}x,\ \beta_t^2) \sim e^{-t}x + \beta_t \mathcal{N}(0, 1)$$

where $\beta_t = \sqrt{1 - e^{-2t}}$. This means that

$$P_t f(x) := \mathbf{E}\left[f(X(t, x))\right] = \int_{\mathbf{R}} f(e^{-t}x + \beta_t y)\mathrm{d}v(y).$$

For $f \in L^1(\mathbf{R} \to \mathbf{R};\ v)$, the dominated convergence theorem entails that

$$P_t f(x) \xrightarrow{t \to \infty} \int_{\mathbf{R}} f(y)\mathrm{d}v(y), \tag{6.5}$$

and the invariance by rotation of the Gaussian distribution implies (as in Lemma 3.3) that

$$X(0, x) \sim \mathcal{N}(0, 1) \implies X(t, x) \sim \mathcal{N}(0, 1),$$

i.e., the Gaussian measure is the stationary and invariant measure of the Markov process X. This can be written as

$$\int_{\mathbf{R}} P_t f(x) d\nu(x) = \int_{\mathbf{R}} f(y) d\nu(y). \tag{6.6}$$

The Itô formula says that

$$f(X(t, x)) = f(x) + \int_0^t f'(X(s, x)) dB(s) + \int_0^t (Lf)(X(s, x)) ds,$$

where for $g \in C^2$,

$$Lg(x) = -xg'(x) + g''(x).$$

Hence,

$$P_t f(x) = f(x) + \int_0^t P_s (Lf)(x) ds.$$

Since L and P_t commute, we also have

$$P_t f(x) = f(x) + \int_0^t P_s Lf(x) ds. \tag{6.7}$$

The operator L^0 has two fundamental properties. By differentiation with respect to t in (6.6), we have

$$\int_{\mathbf{R}} Lf(x) d\nu(x) = \frac{d}{dt} \int_{\mathbf{R}} P_t f(x) d\nu(x) \Big|_{t=0} = 0.$$

Furthermore, a straightforward computation also shows that L is a self-adjoint operator

$$\int_{\mathbf{R}} g(x) Lf(x) d\nu(x) = \int_{\mathbf{R}} f(x) Lg(x) d\nu(x).$$

Example (Gaussian Measure on $\mathbf{R}^n$) If ν is the standard Gaussian measure on $E = \mathbf{R}^n$, all the definitions given in dimension 1 are translated straightforwardly:

$$P_t f(x) = \int_{\mathbf{R}^n} f(e^{-t} x + \beta_t y) d\nu_0(y)$$
$$Lf(x) = -\langle x, Df(x) \rangle_{\mathbf{R}^n} + \Delta f(x)$$

where D is the gradient operator in $\mathbf{R}^n$. The Ornstein–Uhlenbeck is the $\mathbf{R}^n$ valued process whose components are independent one dimensional O–U processes. We finally have

$$\mathcal{E}(f, f) = \int_{\mathbf{R}^n} \langle Df(x), \, Dg(x)\rangle_{\mathbf{R}^n} \mathrm{d}v_0(x)$$

$$\Gamma(f, g)(x) = \langle Df(x), \, Dg(x)\rangle_{\mathbf{R}^n}.$$

Example (Wiener Measure on W*)* If $E = W$, one of our Wiener spaces, and v is the Wiener measure, the situation is much more cumbersome. It is easy to define the semi-group by the Mehler formula (3.28). The Markov process has been identified in (3.33) of Problem 3.1. If we want to generalize formally the definition of L given in $\mathbf{R}^n$, this yields to consider $\langle \omega, \nabla f(\omega)\rangle$ where ω belongs to W and $\nabla f(\omega)$ belongs to $\mathcal{H}$: two spaces that are not in duality. Even worse, the definition of the Laplacian that is the trace of the second order gradient is meaningful only if $\nabla^{(2)} F$ is viewed as an element of $\mathcal{H} \otimes \mathcal{H}$ since the notion of trace does not exist for a map in a Banach space.

The next theorem is far from being trivial and can be found in [9]:

Theorem 6.5 *If* $F \in \mathrm{Lip}_1(W, \; \| \; \|_W)$, *then for any* $t > 0$

$$\frac{\mathrm{d}}{\mathrm{d}t} P_t f(\omega) = \langle \omega, \, \nabla P_t f(\omega)\rangle_{W, W^*} - \mathrm{trace}(\nabla^{(2)} P_t f(\omega)).$$

A remarkably efficient way to construct a Dirichlet structure is to have at our disposal a Malliavin gradient D and to set $L^0 = -D^* D$ where D^* is the adjoint of the gradient.

Example (Gaussian Measures) On $\mathbf{R}$, a standard integration by parts shows that

$$\int_{\mathbf{R}} f'(x)g(x)\mathrm{d}v(x) = \int_{R} f(x)\delta g(x)\mathrm{d}v(x)$$

where δg is given by

$$\delta g(x) = xg(x) - g'(x).$$

Hence, we retrieve that $L = -\delta\nabla$. The same approach works on $\mathbf{R}^n$. On W, we know that $L = \delta\nabla$ (note the harmless change of convention for the sign in front of $\delta\nabla$) by its operation on the chaos, see Theorem 3.10. We also know from Theorem 3.12 that the Mehler formula still holds with this definition of L, and thus Theorem 6.5 is still valid in this presentation.

We are not limited to Gaussian measures. The other nice structures are those related to the Poisson distribution.

*Example (Poisson Distribution on **N** of Parameter ρ)* The space E is **N**, and the gradient is defined by

$$Df(n) = f(n+1) - f(n).$$

The Ornstein–Uhlenbeck process is the process defined in the M/M/∞ queue, see page 129, whose generator is

$$Lf(n) = \rho\big(f(n+1) - f(n)\big) + n\big(f(n-1) - f(n)\big).$$

Example (Poisson Process) For instance, when E is the space of configurations on the compact set K and ν_0 is the distribution of the Poisson process of intensity measure σ, the process X^0 is nothing but the Glauber process G and L^0 is $\mathcal{L}$, see Sect. 5.4.2. The covariance identity of Theorem 5.14 is actually the integration by parts of Theorem 6.4.

The Dirichlet structure may also be useful on the target space as it characterizes the measure μ as the invariant measure of a Markov process $X^\dagger$ of generator $L^\dagger$ and semi-group $P^\dagger$. Remark that for the two main examples, the generator is the sum of two antagonistic parts, which explain the existence of the stationary measure. With this decomposition, the identity $\mathbf{E}[LF] = 0$ is equivalent to the integration by parts formula in the sense of Malliavin calculus.

Example (Poisson Process) The Glauber process contains a part where an atom is added anywhere according to σ and another part that removes one of the atoms, so that the number of atoms does not diverge.

*Example (Wiener Measure on **R**^{n})* The diffusion part (i.e., the Laplacian in L of the Brownian motion part in the differential equation defining X) pushes the process anywhere far from 0; meanwhile, the retraction force (the term in $xf'(x)$) brings X back to the origin.

The main formula for us is the following:

$$\int_{\mathbf{R}} f(y)\mathrm{d}\mu(y) - f(x) = -\int_0^\infty L^\dagger P_t^\dagger f(x)\mathrm{d}t. \tag{6.8}$$

In other presentations of the Stein method, the function

$$f^\dagger : x \longmapsto \int_0^\infty P_t^\dagger f(x)\mathrm{d}t$$

is called the *solution of the Stein equation*. Thus we have

$$\sup_{f \in \mathfrak{F}} \int_{\mathbf{R}} f(y)\mathrm{d}\mu(y) - \int_{\mathbf{R}} f\mathrm{d}\nu = \sup_{f^\dagger} \int_{\mathbf{R}} L^\dagger f^\dagger(x)\mathrm{d}\nu(x). \tag{6.9}$$

6.2 Fourth Order Moment Theorem

The fourth order moment theorem says that a sequence of elements of given Wiener chaos may converge in distribution to the standard Gaussian law provided that the sequences of the fourth moments converge to 3, which is the fourth moment of $\mathcal{N}(0, 1)$.

The target distribution is the usual $\mathcal{N}(0, 1)$ so that

$$L^\dagger f(x) = x f'(x) - f''(x).$$

The initial space is W equipped with Wiener measure, and L^0 is defined by its expression on the chaos. Since

$$\Gamma^0(V, V) = \langle \nabla V, \nabla V \rangle_{\mathcal{H}},$$

we have the following identity:

$$\Gamma^0(\psi(V), \varphi(V)) = \psi'(V)\varphi'(V)\,\Gamma^0(V, V). \tag{6.10}$$

Chain Rule and Poisson Point Process
This last formula no longer holds for the Poisson point process as the gradient does not satisfy the chain rule.

Lemma 6.1 *For $f \in \mathrm{Lip}_1(\mathbf{R}, |\,|)$, let*

$$f^\dagger(x) = \int_0^\infty P_t f(x)\,dt.$$

Then, $f^\dagger$ is twice differentiable and

$$\|(f^\dagger)''\|_\infty \le \sqrt{\frac{2}{\pi}}.$$

Proof Since f is Lipschitz continuous, it is almost everywhere differentiable with a derivative essentially bounded by 1. By the dominated convergence theorem, we get that $P_t f$ is once differentiable with

$$(P_t f)'(x) = e^{-t} \int_{\mathbf{R}} f'(e^{-t}x + \beta_t y)\,d\mu(y)$$

where μ is the standard Gaussian measure on $\mathbf{R}$. Mimicking the proof of Theorem 3.14, we get that $(P_t f)'$ is once differentiable with derivative given by

$$(P_t f)''(x) = \frac{e^{-2t}}{\beta_t} \int_{\mathbf{R}} f'(e^{-t} x + \beta_t y)\, y d\mu(y).$$

Since $\|f'\|_\infty \le 1$, we have

$$\|(f^\dagger)''\|_\infty \le \int_0^\infty \frac{e^{-2t}}{\beta_t} dt \int_{\mathbf{R}} |y| d\mu(y)$$

$$= 1 \times \sqrt{\frac{2}{\pi}}.$$

$\square$

Theorem 6.6 *Let $V \in L^2(E \to \mathbf{R}; \nu_0)$ such that $\mathbf{E}[V] = 0$ and $\mathbf{E}[V^2] = 1$. Then,*

$$\mathrm{dist}_{KR}(V, \mathcal{N}(0, 1)) \le \sqrt{\frac{2}{\pi}} \left| \mathbf{E}\left[\Gamma^0\big((L^0)^{-1}(V), V\big) + 1 \right] \right|.$$

Proof We have to estimate

$$\sup_{f^\dagger:\, f \in \mathrm{Lip}_1(\mathbf{R},|\ |)} \mathbf{E}\left[L^\dagger f^\dagger(V) \right] = \sup_{f^\dagger:\, f \in \mathrm{Lip}_1(\mathbf{R},|\ |)} \mathbf{E}\left[V\, (f^\dagger)'(V) - (f^\dagger)''(V) \right].$$

The Trick: $LL^{-1} = \mathrm{Id}$

In view of this identity and of (6.10) and (6.3), we get

$$\mathbf{E}\left[V\, (f^\dagger)'(V) \right] = \mathbf{E}\left[L^0 (L^0)^{-1} V\, (f^\dagger)'(V) \right]$$

$$= -\mathbf{E}\left[\Gamma^0\big((L^0)^{-1} V,\, (f^\dagger)'(V)\big) \right]$$

$$= -\mathbf{E}\left[(f^\dagger)''(V)\, \Gamma^0\big((L^0)^{-1} V,\, V\big) \right].$$

The result follows from Lemma 6.1.

$\square$

If V belongs to the p-th chaos, $(L^0)^{-1} V = p^{-1} V$; thus we get

$$\mathrm{dist}_{KR}(V, \mathcal{N}(0, 1)) \le \frac{1}{p}\sqrt{\frac{2}{\pi}} \left| \mathbf{E}\left[\Gamma^0(V, V) + p \right] \right|. \tag{6.11}$$

We then estimate the right-hand side of (6.11) by computing the variance of $\Gamma^0(V, V)$. This requires two technical results.

Theorem 6.7 *Let* $V \in \oplus_{k=0}^p \mathcal{C}_k$. *Then, for any* $\eta \geq p$,

$$\mathbf{E}\left[V(L^0 - \eta\,\mathrm{Id})^2 V\right] \leq -\eta\,\mathbf{E}\left[V(L^0 - \eta\,\mathrm{Id})V\right] \leq \eta c\,\mathbf{E}\left[V(L^0 - \eta\,\mathrm{Id})^2 V\right],$$
(6.12)

where

$$c = \frac{1}{\eta - p} \wedge 1.$$

Proof **Step 1**   Since V belongs to $\oplus_{k=0}^p \mathcal{C}_k$, we can write

$$V = \sum_{k=0}^p J_k^s(\dot{v}_k) \text{ and } L^0 V = \sum_{k=0}^p k\, J_k^s(\dot{v}_k)$$
(6.13)

It follows that

$$\mathbf{E}\left[V(L^0 - \eta\,\mathrm{Id})^2 V\right] = \mathbf{E}\left[V L^0 (L^0 - \eta\,\mathrm{Id})V\right] - \eta\mathbf{E}\left[V(L^0 - \eta\,\mathrm{Id})V\right]$$

$$= \mathbf{E}\left[V\sum_{k=0}^p k(k - \eta)\, J_k^s(\dot{v}_k)\right] - \eta\mathbf{E}\left[V(L^0 - \eta\,\mathrm{Id})V\right].$$

By orthogonality of the chaos,

$$\mathbf{E}\left[V\sum_{k=0}^p k(k - \eta)\, J_k^s(\dot{v}_k)\right] = \sum_{k=0}^p k(k - \eta)\mathbf{E}\left[J_k^s(\dot{v}_k)^2\right] \leq 0,$$

in view of the assumption on η. The first inequality follows.

Step 2   Following the same lines of thought,

$$-\mathbf{E}\left[V(L^0 - \eta\,\mathrm{Id})V\right] = \sum_{k=0}^p (\eta - k)\mathbf{E}\left[J_k^s(\dot{v}_k)^2\right]$$

$$\leq c\sum_{k=0}^p (\eta - k)^2\mathbf{E}\left[J_k^s(\dot{v}_k)^2\right]$$

$$= c\mathbf{E}\left[V(L^0 - \eta\,\mathrm{Id})^2 V\right].$$

The proof is thus complete.

$\square$

Remark 6.4 Note that the proof requires V to belong to a finite sum of chaos to choose a finite η.

Lemma 6.2 *Let $V \in \mathfrak{C}_p$ and Q a polynomial of degree two. Then,*

$$\mathbf{E}\left[Q(V)(L^0 + ap\,\mathrm{Id})Q(V)\right] = p\mathbf{E}\left[aQ^2(V) - \frac{Q'(V)^3\,V}{2Q''(V)}\right]. \tag{6.14}$$

Proof Apply (6.3) and (6.10) to obtain

$$\mathbf{E}\left[Q(V)\,L^0Q(V)\right] = -\mathbf{E}\left[\Gamma(Q(V))\right]$$
$$= -\mathbf{E}\left[Q'(V)^2\,\Gamma(V)\right].$$

Since $Q^{(3)} = 0$, we have

$$\left(\frac{Q'(X)^3}{3Q''(X)}\right)' = \frac{3Q'(X)^2Q''(X)^2}{3Q''(X)^2} = Q'(X)^2,$$

so that in view of (6.10), we get

$$\mathbf{E}\left[Q(V)\,L^0Q(V)\right] = -\mathbf{E}\left[\Gamma\left(\frac{Q'(V)^3}{3\,Q''(V)},\, V\right)\right]$$
$$= -\mathbf{E}\left[\frac{Q'(V)^3}{3\,Q''(V)}\,L^0V\right]$$
$$= -p\mathbf{E}\left[\frac{Q'(V)^3}{3\,Q''(V)}\,V\right],$$

thanks again to (6.3).   □

Theorem 6.8 *For $V \in \mathfrak{C}_p$, we have*

$$\mathbf{E}\left[(\Gamma(V) + p)^2\right] \le \frac{p^2}{6}\left(\mathbf{E}\left[V^4\right] - 6\mathbf{E}\left[V^2\right] + 3\right).$$

Proof **Step 1** By the very definition of Γ, for $V \in \mathfrak{C}_p$, we have

$$\Gamma(V) + p = \frac{1}{2}L^0(V^2) - VL^0V + p = \frac{1}{2}L^0(V^2) - pV^2 + p$$

and

$$\frac{1}{2}(L^0 - 2p\,\mathrm{Id})(V^2 - 1) = \frac{1}{2}L^0(V^2) - pV^2 + p.$$

Step 2 It follows that

$$\mathbf{E}\left[\left(\Gamma(V) + p\right)^2\right] = \frac{1}{4}\mathbf{E}\left[\left((L^0 - 2p\,\mathrm{Id})\mathfrak{H}_2(V, 1)\right)^2\right].$$

Since L^0 is a self-adjoint operator, this yields

$$\mathbf{E}\left[\left(\Gamma(V) + p\right)^2\right] = \frac{1}{4}\mathbf{E}\left[\mathfrak{H}_2(V, 1)\,(L^0 - 2p\,\mathrm{Id})^2\mathfrak{H}_2(V, 1)\right].$$

Step 3 The formula for the product of iterated integrals (3.24) implies that $V^2 \in \bigoplus_{k=0}^{2p}\mathfrak{C}_k$; hence, we are in position to apply Theorem 6.7 with $\eta = p$:

$$\mathbf{E}\left[\left(\Gamma(V) + p\right)^2\right] \leq \frac{p}{4}\mathbf{E}\left[\mathfrak{H}_2(V)\,(L^0 - 2p\,\mathrm{Id})\mathfrak{H}_2(V)\right].$$

According to Lemma 6.2 with $a = 2$, we obtain

$$\mathbf{E}\left[\left(\Gamma(V) + p\right)^2\right] \leq \frac{p^2}{4}\mathbf{E}\left[2\mathfrak{H}_2(V) - \frac{V\mathfrak{H}_2'(V)^3}{3\mathfrak{H}_2''(V)}\right]$$

$$= \frac{p^2}{4}\mathbf{E}\left[2(V^2 - 1)^2 - \frac{4}{3}V^4\right]$$

$$= \frac{p^2}{6}\left(\mathbf{E}\left[V^4\right] - 6\mathbf{E}\left[V^2\right] + 3\right).$$

The proof is thus complete.

□

Corollary 6.1 *For $V \in \mathfrak{C}_p$,*

$$\mathrm{dist}_{\mathrm{KR}}\left(V, \mathcal{N}(0, 1)\right) \leq \frac{1}{\sqrt{3\pi}}\left(\mathbf{E}\left[V^4\right] - 6\mathbf{E}\left[V^2\right] + 3\right)^{1/2}.$$

Proof Combine (6.11), Cauchy–Schwarz inequality, and Theorem 6.8.   □

6.3   Poisson Process Approximation

Convergence toward the Gaussian measure is not the whole story. We can also investigate the distance between point processes. The basic formula is, as usual, the integration by parts formula (see Theorem 5.9). This gives the Dirichlet structure on the target structure. When the initial probability space is also a configuration space, the so-called GNZ formula (for Georgii–Nguyen–Zessin) is in fact an integration by parts formula.

Definition 6.3 The set of finite configurations is denoted by $\mathfrak{N}_E^f$. It can be decomposed as the disjoint union of the $\mathfrak{N}_E^{(n)}$ where

$$\mathfrak{N}_E^{(n)} = \{\phi \in \mathfrak{N}_E, \ \phi(E) = n\}.$$

Intuitively, a configuration with n points is an element of E^n, but since there is no privileged order in the enumeration of the elements of a configuration, we must identify all the n-tuples of E^n that differ only by the order of their elements. Mathematically speaking, this amount to consider the quotient space $E_s^n = E^n/\mathfrak{S}_n$, where $\mathfrak{S}_n$ is the group of permutations over $\{1, \cdots, n\}$: Two elements $x = (x_1, \cdots, x_n)$ and $y = (y_1, \cdots, y_n)$ are in relation if there exists $\sigma \in \mathfrak{S}_n$ such that:

$$\left(y_1, \cdots, y_n\right) = \left(x_{\sigma(1)}, \cdots, x_{\sigma(n)}\right).$$

The set E_s^n is the set of all equivalence classes for this relation.

We thus have a bijection $\mathfrak{c}_n$ between $\mathfrak{N}_E^{(n)}$ and E_s^n. A function F defined on $\mathfrak{N}_E^{(n)}$ can be transferred to a function defined on E_s^n, but it is more convenient to see it as a function on E^n with the additional constraint to be symmetric.

Definition 6.4 For $\phi = \{x_1, \cdots, x_n\} \in \mathfrak{N}_E^{(n)}$, let $x = (x_1, \cdots, x_n) \in E^n$ and $\mathfrak{p}_n(x)$ the equivalence class of x in E_s^n. Let F be measurable from $\mathfrak{N}_E^{(n)}$ to $\mathbf{R}$, and define $\tilde{F} : E^n \to \mathbf{R}$ by

$$\tilde{F}(x_1, \cdots, x_n) = F\left(\mathfrak{c}_n^{-1}\left(\mathfrak{p}_n(x_1, \cdots, x_n)\right)\right).$$

By its very definition, $\tilde{F}$ is symmetric. For the sake of simplicity, we again abuse a notation and write F instead of $\tilde{F}$.

Definition 6.5 For σ a reference measure on E, N admits Janossy densities $(j_n, \ n \geq 0)$ if we can write for any $F \in L^\infty(\mathfrak{N}_E^f \to \mathbf{R}; \mathbf{P})$,

$$\mathbf{E}[F(N)] = F(\emptyset)P(N = \emptyset)$$

$$+ \sum_{n=1}^{\infty} \frac{1}{n!} \int_{E^k} F(x_1, \cdots, x_n) \, j_n(x_1, \cdots, x_n) \, \mathrm{d}\sigma(x_1) \ldots \mathrm{d}\sigma(x_n).$$

The Janossy density j_n is intuitively defined as the probability to have exactly n atoms and that those atoms are located in the vicinity of $(x_1, \cdots, x_n)$.

By the very construction of the Poisson point process, the Janossy densities of a Poisson point process are easy to calculate.

Corollary 6.2 *Let N be a Poisson point process of intensity σ. For any bounded $F : \mathfrak{N}_E^f \to \mathbf{R}$,*

$$\mathbf{E}[F(N)] = e^{-\sigma(E)} F(\emptyset)$$

$$+ e^{-\sigma(E)} \sum_{n=1}^{\infty} \frac{1}{n!} \int_{E^n} F(x_1, \cdots, x_n) \, d\sigma(x_1) \ldots d\sigma(x_n). \qquad (6.15)$$

This means that N admits Janossy densities:

$$j_n(x_1, \cdots, x_n) = e^{-\sigma(E)}.$$

Proof Since $\sigma(E)$ is finite, we can write

$$\mathbf{E}[F(N)] = \sum_{n=0}^{\infty} \mathbf{E}[F(N) \mid N(E) = n] \, \mathbf{P}(N(E) = n).$$

According to the construction of the Poisson point process, given $N(E) = n$, the distribution of the atoms $(X_1, \cdots, X_n)$ of N is $(\sigma(E)^{-1}\sigma)^{\otimes n}$. This means that for $n > 0$,

$$\mathbf{E}[F(N) \mid N(E) = n] = \frac{1}{\sigma(E)^n} \int_{E^n} F(x_1, \cdots, x_n) \, d\sigma(x_1) \ldots d\sigma(x_n).$$

For $n = 0$, it is a tautology to say that $F(N) = F(\emptyset)$. Hence,

$$\mathbf{E}[F(N)] = F(\emptyset) e^{-\sigma(E)}$$

$$+ e^{-\sigma(E)} \sum_{n=1}^{\infty} \frac{\sigma(E)^n}{n!} \frac{1}{\sigma(E)^n} \int_{E^n} F(x_1, \cdots, x_n) \, d\sigma(x_1) \ldots d\sigma(x_n).$$

The proof is thus complete. $\qquad \square$

Example (Janossy Densities of Poisson Process) A Poisson point process N of intensity σ is a finite point process if and only $\sigma(E) < \infty$. Then (6.15) induces that

$$j_n(x_1, \cdots, x_n) = e^{-\sigma(E)}.$$

Definition 6.6 [5, Section 15.5] If N is a finite point process with Janossy measures $(j_n, n \geq 0)$, we define the Papangelou intensity by

$$c(\{x_1, \cdots, x_n\}, x) = \frac{j_{n+1}(x_1, \cdots, x_n, x)}{j_n(x_1, \cdots, x_n)}.$$

The quantity $c(N, x)$ can be seen intuitively as the probability to have a particle at x given the observation N.

We then have the so-called GNZ (for Georgii–Nguyen–Zessin) formula:

Theorem 6.9 *For any bounded* $U : \mathfrak{N}_E \times E \to \mathbf{R}$, *we have*

$$\mathbf{E}\left[\int_E U(N \ominus x, \, x) \, dN(x)\right] = \mathbf{E}\left[\int_E U(N, \, x) \, c(N, x) \, d\sigma(x)\right]. \qquad (6.16)$$

Proof **Step 1** Remark that

$$\int_E U(N \ominus x, \, x) dN(x)\bigg|_{N=\emptyset} = 0.$$

Hence, according to the definition of Janossy densities, we have

$$\mathbf{E}\left[\int_E U(N \ominus x, \, x) dN(x)\right]$$

$$= \sum_{k=1}^{\infty} \frac{1}{k!} \int_{E^k} \left(\sum_{j=1}^{k} U(\{x_1, \cdots, x_k\}, x_j)\right) j_k(x_1, \cdots, x_k) d\sigma^{\otimes k}(x).$$

Step 2 It is clear that for any $j \in \{1, \cdots, k\}$,

$$\int_{E^k} U(\{x_1, \cdots, x_k\}, x_j) d\sigma^{\otimes k}(x) = \int_{E^k} U(\{x_1, \cdots, x_k\}, x_k) d\sigma^{\otimes k}(x).$$

Hence,

$$\mathbf{E}\left[\int_E U(N \ominus x, \, x) dN(x)\right]$$

$$= \sum_{k=1}^{\infty} \frac{k}{k!} \int_{E^k} U(\{x_1, \cdots, x_k\}, x_k) j_k(x_1, \cdots, x_k) d\sigma^{\otimes k}(x).$$

Step 3 The definition of c can be read as

$$c(\{x_1, \cdots, x_{k-1}\}, x_k) \, j_{k-1}(x_1, \cdots, x_{k-1}) = j_k(x_1, \cdots, x_k),$$

from which we derive

$$\mathbf{E}\left[\int_E U(N \ominus x, x)\mathrm{d}N(x)\right]$$

$$= \sum_{k=1}^{\infty} \frac{1}{(k-1)!} \int_{E^{k-1}} \left(\int_E U(\{x_1, \cdots, x_k\}, x_k)c(\{x_1, \cdots, x_{k-1}\}, x_k)\,\mathrm{d}\sigma(x_k)\right)$$

$$\times j_{k-1}(x_1, \cdots, x_{k-1})\,\mathrm{d}\sigma^{\otimes(k-1)}(x).$$

Apply once more the definition of the Janossy densities to obtain the right-hand side of (6.16).

□

An immediate corollary of (6.16) is the following:

Corollary 6.3 (Integration by Parts Formula for General Point Processes) *Let N be a finite point process of Papangelou intensity c. For any $F \in L^{\infty}(\mathfrak{N}_E \to \mathbf{R}; \mathbf{P})$ and any $U \in L^{\infty}(\mathfrak{N}_E \times E \to \mathbf{R}; \mathbf{P} \otimes \sigma)$, the following identity holds*

$$\mathbf{E}\left[F(N) \left(\int_E U(N \ominus y, y)\mathrm{d}N(y) - \int_E U(N, y)\,c(N, y)\mathrm{d}\sigma(y)\right)\right]$$

$$= \mathbf{E}\left[\int_E D_y F(N)\, U(N, y)c(N, y)\mathrm{d}\sigma(y)\right],$$

where D is defined as before by

$$D_y F(N) = F(N \oplus y) - F(N).$$

We can then say that

$$\delta U(N) = \int_E U(N \ominus y, y)\mathrm{d}N(y) - \int_E U(N, y)\,c(N, y)\mathrm{d}\sigma(y).$$

Remark 6.5 As for a Poisson process, the Janossy densities are all equal to $e^{-\sigma(E)}$, and the Papangelou intensity is equal to 1. In this setting, the GNZ formula reduces to the Campbell–Mecke formula (5.11).

If $\mathrm{d}\sigma(x) = m(x)\mathrm{d}\ell(x)$, then it is customary to take as reference measure $\mathbf{P} \otimes \ell$ so that the Papangelou intensity becomes $c(N, x) = m(x)$.

We can now state the main theorem that bounds the distance between a point process described by its Papangelou intensity and a Poisson point process of intensity $m\,\mathrm{d}\ell$ where

$$m(x) = \mathbf{E}[c(N, x)]. \tag{6.17}$$

Definition 6.7 The space of test functions is the set of Lipschitz functions in the sense

$$|F(N \oplus x) - F(N)| \le 1, \ \forall x \in E.$$

It is denoted by $\mathrm{Lip}_1(\mathfrak{N}_E, \mathrm{dist}_{\mathrm{TV}})$.

Theorem 6.10 *Let M be a point process of Papangelou intensity c with respect to the measure $\mathbf{P} \otimes \ell$ and m defined by (6.17). Let π^σ be the distribution of the Poisson point process of intensity $d\sigma(x) = m(x)d\ell(x)$. Assume that $\sigma(E) < \infty$. Then,*

$$\sup_{F \in \mathrm{Lip}(\mathfrak{N}_E)} \left(\mathbf{E}\left[F(M)\right] - \int_{\mathfrak{N}_E} F \, d\pi^\sigma \right) \le \mathbf{E}\left[\int_E |c(M,x) - m(x)| d\ell(x) \right].$$

Proof According to the construction of the Glauber process, we have

$$\int_{\mathfrak{N}_E} F \, d\pi^\sigma - F(M) = \int_0^\infty L P_t F(M) \, dt$$

where

$$-LF(M) = \int_E \Big(F(M \oplus x) - F(M) \Big) d\sigma(x) + \int_E \Big(F(M \ominus x) - F(M) \Big) dM(x).$$

In view of Corollary 6.3 with $F = 1$, we have

$$\mathbf{E}\left[\int_E \Big(F(M \ominus x) - F(M) \Big) dM(x) \right]$$
$$= \mathbf{E}\left[\int_E \Big(F(M) - F(M \oplus x) \Big) \right] c(M,x) \, d\ell(x).$$

Thus,

$$\int_{\mathfrak{N}_E} F \, d\pi^\sigma - \mathbf{E}\left[F(M)\right] = \mathbf{E}\left[\int_0^\infty \int_E D_x P_t F(M) \Big(m(x) - c(M,x) \Big) d\ell(x) \, dt \right].$$

Moreover, (5.21) entails that $D_x P_t F(M) = e^{-t} P_t D_x F(M)$. Recall that F is Lipschitz; hence, $|D_x F| \le 1$ for all $x \in E$, and (5.18) entails that

$$|P_t D_x F| \le \mathbf{E}[\mathbf{1}] = 1.$$

Thus we have

$$\mathbf{E}\left[F(N^\sigma)\right] - \mathbf{E}\left[F(M)\right] \le \int_0^\infty e^{-t} \, dt \times \int_E |m(x) - c(M,x)| \, d\ell(x).$$

The result follows.   □

6.4 Problems

6.1 A point process M is a Gibbs point process on $E = \mathbf{R}^k$, of order 2 and temperature $\beta > 0$ if its Janossy densities (with respect to the Lebesgue measure) are given by

$$j_n(x_1, \cdots, x_n) = \exp\left(-\beta \sum_{j=1}^{n} \psi_1(x_j) - \beta \sum_{1 \le i < j \le n} \psi_2(x_i, x_j)\right)$$

where ψ_1 and ψ_2 are two non-negative functions on E and $E \times E$, respectively, such that ψ_2 is bounded, symmetric, and

$$\int_E e^{-\beta \psi_1(x)} \, d\ell(x) < \infty.$$

1. With the non-negativity of ψ_2, show that

$$\mathbf{E}[M(E)] \le \int_E e^{-\beta \psi_1(x)} \, d\ell(x).$$

2. Show that the Papangelou intensity of M is given by

$$c(M, x) = \exp\left(-\beta \psi_1(x) - \beta \int_E \psi_2(x, y) \, dM(y)\right).$$

3. Show that

$$|c(M, x) - e^{-\beta \psi_1(x)}| \le \beta e^{-\beta \psi_1(x)} \|\psi_2\|_\infty M(E).$$

Indication: remember that for $x \ge 0$, $1 - e^{-x} \le x$.

4. For $d\sigma(x) = \exp(-\beta \psi_1(x)) \, d\ell(x)$, show that

$$\sup_{F \in \mathrm{Lip}(\mathfrak{N}_E)} \left(\mathbf{E}[F(M)] - \int_{\mathfrak{N}_E} F \, d\pi^\sigma\right) \le \beta \|\psi_2\|_\infty \left(\int_E e^{-\beta \psi_1(x)} \, d\ell(x)\right)^2.$$

6.2 (Superposition of Weakly Repulsive Processes) A common interpretation of the Papangelou intensity is to say that $c(\phi, x)$ represents the infinitesimal probability to have an atom at position x given the observation ϕ. Thus, a possible definition of repulsiveness could be to impose that

$$\phi \subset \psi \implies c(\psi, x) \le c(\phi, x).$$

We here define a less restrictive notion of *weak* repulsiveness:

$$c(\phi, x) \le c(\emptyset, x), \ \forall x \in E, \ \forall \phi \in \mathfrak{N}_E.$$

Let $p_0 = \mathbf{P}(M(\mathbf{R}^d) = 0)$. Assume that M is weakly repulsive.

1. Using the GNZ formula, show that

$$p_0 \sigma(x) \le p_0 c(\emptyset, x).$$

2. Show that

$$\sigma(x) \ge p_0 c(\emptyset, x).$$

3. Derive that

$$|c(\emptyset, x) - \sigma(x)| \le (1 - p_0) c(\emptyset, x).$$

6.5   Notes and Comments

The introduction of distances between probability is mainly inspired by Dudley [6, 7], Villani [10]. The Stein method dates back to the seventies when it was created by C. Stein for the convergence toward the one dimensional Gaussian standard distribution. It was quickly extended to the convergence toward the Poisson distribution (see [3] and references therein for a more complete history). The principle of the method is always the same, but the computations are adhoc to each situation, so it has yielded a vast number of papers during the last thirty years. The papers by Nourdin and Peccati who introduced the Malliavin calculus in this framework renewed the interest and the scope of the method (see [8]), see [2] for a recent and thorough survey. The most striking result was the fourth moment theorem the proof of which has been recently greatly simplified in [1]. We followed this line of thought in Sect. 6.2.

The links between Malliavin calculus and Dirichlet forms are the core of [4].

References

1. E. Azmoodeh, S. Campese, G. Poly, Fourth moment theorems for Markov diffusion generators. J. Funct. Anal. **266**(4), 2341–2359 (2014)
2. E. Azmoodeh, G. Peccati, X. Yang, Malliavin–Stein method: a survey of some recent developments. Mod. Stoch. Theory Appl. **8**, 141–177 (2021)
3. A.D. Barbour, L.H.Y. Chen, *An Introduction to Stein's Method*. Lecture Notes Series, vol. 4 (National University of Singapore, Singapore, 2005)

4. N. Bouleau, F. Hirsch, *Dirichlet Forms and Analysis on Wiener Space*, vol. 14 (Walter de Gruyter & Co., Berlin, 1991)
5. D.J. Daley, D. Vere-Jones, *An Introduction to the Theory of Point Processes. Vol. II*. Springer Series in Statistics (Springer, New York, 1988)
6. R.M. Dudley, Distances of probability measures and random variables. Ann. Math. Stat. **39**(5), 1563–1572 (1968)
7. R.M. Dudley, *Real Analysis and Probability*, vol. 74 (Cambridge University Press, Cambridge, 2002)
8. I. Nourdin, G. Peccati, *Normal Approximations with Malliavin Calculus: From Stein's Method to Universality*. (Cambridge University Press, Cambridge, 2012)
9. H.-H. Shih, On Stein's method for infinite-dimensional Gaussian approximation in abstract Wiener spaces. J. Funct. Analy. **261**(5), 1236–1283 (2011)
10. C. Villani, *Topics in Optimal Transportation*. Graduate Studies in Mathematics, vol. 58. (American Mathematical Society, Providence, 2003)

Index

© The Author(s), under exclusive license to Springer Nature Switzerland AG 2022
L. Decreusefond, *Selected Topics in Malliavin Calculus*,
Bocconi & Springer Series 10, https://doi.org/10.1007/978-3-031-01311-9

Printed in the United States
by Baker & Taylor Publisher Services